POSTHUMAN PLANTS

Rethinking the Vegetal through Culture, Art, and Poetry

John Charles Ryan

POSTHUMAN PLANTS

Rethinking the Vegetal through Culture, Art, and Poetry

John Charles Ryan

COMMON GROUND PUBLISHING 2015

First published in 2015 in Champaign, Illinois, USA
by Common Ground Publishing LLC
as part of On Sustainability books

Library of Congress Cataloging-in-Publication Data

Names: Ryan, John (John Charles) (Poet), author.
Title: Posthuman plants : rethinking the vegetal through culture, art, and
   poetry / John Charles Ryan.
Other titles: Rethinking the vegetal through culture, art, and poetry
Description: Champaign, Illinois : Common Ground Publishing, 2015. | Includes
   bibliographical references and index. | Description based on print version
   record and CIP data provided by publisher; resource not viewed.
Identifiers: LCCN 2015040769 (print) | LCCN 2015037122 (ebook) | ISBN
   9781612298221 (pbk) | ISBN 9781612298238 (pdf) | ISBN 9781612298221 (pbk :
   alk. paper) | ISBN 9781612298238 (pdf : alk. paper)
Subjects: LCSH: Ethnobotany--Australia--Western Australia. | Human-plant
   relationships. | Botany in literature. | Biogeography.
Classification: LCC GN476.73 (print) | LCC GN476.73 .R93 2015 (ebook) | DDC
   578.09--dc23
LC record available at http://lccn.loc.gov/2015040769

Cover Photo Credit: Nikki Green, "Interconnection (Vav)" (2015), image hand printed from carved lino blocks. Flower: WA Smokebush (Conospermum stoechadis)

Table of Contents

ACKNOWLEDGEMENTS

I acknowledge the Nyoongar people – the traditional owners and custodians of the South-West of Western Australia.

I thank the Office of Research and Innovation as well as the interdisciplinary CREATEC research group at Edith Cowan University for an encouraging environment and financial support during the writing of these chapters and the preparation of this book. The most significant support has come in the form of a Postdoctoral Research Fellowship in Communications and Arts (2012–2015) for which I am tremendously grateful to ECU.

I also wish to recognize the heroic efforts of the editors, referees, and proofreaders of the journals in which versions of some of these chapters have appeared, both online and in print. They include *Arc Poetry Magazine*, *Australian Garden History* (Richard Aitken), *Conscious Living Magazine*, *Cordite Poetry Review* (Kent MacCarter), *Global Media Journal* (Myra Gurney), *Griffith Review* (Anna Haebich and Rosemary Stevens), *Humanities* (MDPI), *Media International Australia* (Lelia Green and Sarah Pink), and the *Refereed Proceedings of the Buddhism and Australia Conference 2015* (Marju Broder). Their intellectual generosity has greatly improved this collection. Any error of content, style, or logic in *Posthuman Plants*, however, is entirely mine.

Special thanks go to the Monastery Guesthouse at the New Norcia Benedictine Community for providing a wonderfully meditative space for a self-contained writing retreat during which I completed portions of this book.

Of course, I am grateful for the fruits (abstract or otherwise) of the places that inform *Posthuman Plants* – most conspicuously the South-West of Western Australia where I have lived since 2008. To the poets, thinkers, and activists here, my sincere admiration. May the human (superhuman, unhuman, extrahuman, more-than-human, other-than-human) voices continue to resound, thrive, and inspire.

Prologue

In the founding years of the Swan River Colony of Western Australia, an overland route between Perth and Albany became a preoccupation of the new government under Captain Stirling. A serviceable track between the Swan River and King George Sound would be the basis of a vital economic and communications network linking the isolated settlements. Moreover, a continuous road would serve as a potent symbol of settler ambition in the harsh and unfamiliar Western Australian landscape.

For today's travelers along the Albany Highway, the journey reverberates with intriguing anecdotes from the historical record. On the one hand, surveyors, explorers, convicts, homesteaders, and mail carriers endured thirst, starvation, isolation, and the elements as they crossed the unmapped hinterlands of the colony. On the other, early naturalists and botanical artists experienced a sense of euphoria in encountering profuse wildflowers carpeting the hills and valleys during spring months. And, on closer inspection, the plants themselves still seem to confound the botanical norms of the other hemisphere (Ryan 2012).

The 410-kilometer trip now takes approximately five hours by car. Southbound travelers leave the Perth suburbs of Kelmscott and Armadale, ascend the Darling Scarp, cross the Hotham and Williams rivers and, in time, gain a sudden glimpse of the mountainous Stirling Range and Porongorups to the east before arriving in Albany and the Great Southern region. Not limited to the automobile, twenty-first century sojourners can also still traverse the landscape by coach, train, horseback, or foot.

The pedestrian-only Bibbulmun Track parallels the Albany Highway from Jarrahdale State Forest and the Monadnocks Reserve, intersecting the bitumen at North Bannister, and then striking a southwesterly course through the karri country of Pemberton and Walpole. Bearing the name of the Nyoongar people of the South-West of WA, the 963-kilometer "Bib" track was proposed by bushwalker Geoff Schafer in 1972 and completed nearly twenty-five years later, including trackside shelters and water tanks. Yellow Dreamtime serpent route markers can be found near the Bibbulmun's terminus at the Albany Visitor Center overlooking Princes Royal Harbour at journey's end.

After Armadale, Bannister is the next way mark. In 1830, explorer and pastoralist Thomas Bannister led the first attempt by a European to forge an overland passage. The party paralleled the route of the later highway until crossing the Williams River, followed the Hillman River south, and veered perilously off course toward Mount Roe and the Southern Ocean. On reaching Broke Inlet near present-day Walpole, exhausted, famished, and disoriented, Bannister and his party walked another nineteen days east through thickly forested country until arriving at the Albany outpost on King George Sound (which had been named in 1791 after the reigning monarch by British explorer George Vancouver).

Now only faintly detectable on maps, the towns of Bannister and North Bannister were once crucial resupply points for the fortnightly horse-drawn Royal Mail service. By 1840, mail carriers left Perth on the fifteenth of each month, taking twelve days to reach Albany. After three days recuperation, the team commenced the return trip on the first of the following month. The loss of freight to fires, floods, and bushrangers incited constant public complaints. The horse-drawn mail cycle continued until the advent of railway and road lines connecting the settlements.

A short diversion from the highway leads to Araluen Botanic Park. Sheltered from prevailing winds, the valley microclimate produces a unique botanical haven. The area is colder and wetter than Perth, offering ideal conditions for rhododendrons, azaleas, and species uncommon to the Mediterranean climate of the city. Businessman J.J. "Boss" Simons envisioned a holiday camp where urban youth could return to nature by tending gardens, felling trees, and constructing dams – Araluen literally means "singing waters" and "place of lilies" (Ayris 2001).

Nowhere are these poetic nuances more redolent than at the Grove of the Unforgotten, designed as a memorial to eighty-nine boys killed in World War I. Pencil pines (*Cupressus* spp.) mixed with indigenous marri trees (*Corymbia callophylla*) enclose a series of terraces shaped like a lyre, a symbol of music. Waterfalls topple through the lyre's center to the Pool of Reflection, bestrewn with water lilies – all evocative of the 1930s. Magnolias, fuchsias, leschenaultias, and pimeleas can be found in October and November.

Leaving the Araluen oasis, some of the South-West plants that provoked the curiosity of early diarists, writers, and artists are to be found. Balga or grasstree (*Xanthorrhoea preissii*) is a primordial and profoundly slow-growing tree that can live for over five-hundred years. In fact, there are unconfirmed reports of a grove in Gosnells, at the head of the Albany Highway, exceeding seven hundred years. In its distinct growth habit, the balga symbolizes the adaptability of the biota here. The composer and traveler Thomas Wood wrote of his visit in the bestseller *Cobbers* (1943), personifying the tree as a "strange fascination." "He stands, twisted and knobbly, among the moss and feathery bracken," he wrote, "wearing a mop of tousled grass overtopped by a spear" (Wood 1943).

Although not as conspicuous as the balga, the western quandong (*Santalum acuminatum*), a relative of the aromatic sandalwood that was transported to Albany along this route, is another small tree populating the overland journey. Bannister characterized the quandong, using his powers of sight and taste, as about as large as an English plum tree. "It bears a nut almost round, having a strong shell, and as large as a pigeon's egg, with small holes in it similar to the almond, and an outcovering, which it throws off apparently when ripe," he observed (Bannister 1833, 101). "The kernel we found nutritive, possessing a glutinous property and very easy of digestion" (Bannister 1833, 101). Given the time of year of his overland journey, it is clear that Bannister sampled the dregs of the season's quandong riches. The tangy, crimson-colored drupes ripen earlier, in late spring or summer.

Beginning in early November, the luminous golden flower of the Christmas tree (*Nuytsia floribunda*) emblazons the stretch between the Williams and Arthur River districts (see chapter 7). The peripatetic English botanical artist Marianne North passed through by carriage. In her memoir published in 1892, North wrote in ecstatic terms about the hemi-parasitic tree: "I shall never forget one plain we came to, entirely surrounded by the nuytsia or mistletoe trees, in a full blaze of bloom. It looked like a bush-fire without smoke. The tree are, many of them, as big as average oaks in our hedgerows at home, and the stems are mere pith, not wood" (North 2011, 153). Also in the late nineteenth century, Canadian novelist Gilbert Parker, travelling on the new railway, echoed North's appreciation of the pleasing composition of the bush during this time of year: "the yellow cabbage-tree flower [*Nuytsia*] is gleaming near, flanked by the white-and-green banksia, and a blossoming gum-tree is full of a regal beauty" (Parker 1892, 368).

Between 1835 and 1837, surveyor Alfred Hillman had endured several trips between the settlements of the region. On the first of these, he found the fresh water spring at Kojonup; but on his next trip, he and party nearly perished of thirst near the Beaufort River. Nowadays, the Kojonup area is known for its orchid diversity, including white spider, darting spider, greenhood, rabbit, donkey, and jug orchids. The Kojonup Visitor Centre, including Kodja Place, is also a veritable treasure trove for information on the cultural and natural aspects of the region.

Views of the Stirling Range and its highest peak, Bluff Knoll, soon appear, as surveyor John Septimus Roe exclaimed in a journal entry from November 9,1835, "burst[ing] on our view in great magnificence as we rounded the crest of a clear rising ground, from which the base of the nearer hills appeared to be distant not more than 6 or 7 miles, while the whole extent of their conical and picturesque summits were spread out before us" (Roe 2005, 479). Roe's contemporary, botanist James Drummond, found plants everywhere he could look: "I had scarcely time to make myself acquainted with this fine Banksia when I found another exceedingly interesting and beautiful plant [*Darwinia macrostegia*]," the bracts of which he compared to "the petals of the finest tulip and they are almost as large, hanging in a bell" (quoted in Barker 1996, 5). There are sixty species of *Darwinia* in Western Australia, including the success bell (*Darwinia nubigena*) and many others that only occur there. In fact Drummond established a garden near Perth where he cultivated plants obtained from his inland expeditions, sending seeds and specimens from this stock to London's Kew Gardens.

The "fine Banksia" on Mount Mongerup was described by Drummond as "a splendid *Banksia* [*Banksia solandri*], with leaves more than 9 inches long and about 5 wide, irregularly jagged and serrated like an English Oak" (quoted in Barker 1996, 5). Similarly, on her overland passage, artist Marianne North was enamored of banksias "covered with their young leaves and shoots of rich yellow, brown, or white [...] the native wigwams of bark or leaves looked picturesque under them" (North 2011, 154). The modern-day fascination for the genus is palpable at Banksia Farm in Mount Barker, about thirty minutes by car from Albany. The farm is known for its collection of banksias from around Australia.

Albany had been established in 1826 when three British ships travelled from Sydney to stake claim to the region and discourage French settlement. Naturalists and artists soon followed. In 1854, the renowned phycologist W.H. Harvey arrived here before departing overland for Fremantle. Botanical illustrator Ellis Rowan painted "Verticordia habrantha" in 1885 around the time she befriended Marianne North. It was North who described Albany as "a natural flower-garden," observing that "in one place I sat down, and without moving could pick twenty-five different flowers within reach of my hand" (North 2011, 149). In particular, she extolled the scallop hakea (*Hakea cucullata*) as "one of the remarkable plants of the world" (North 2011, 150) and noted "strange plants known as 'kangaroo's feet' [*Anigozanthos* spp.]" (152).

The chortling cascades of Araluen, the orchids of Kojonup, and the alpine wildflowers of the Stirling Range are a few of the many beguiling features of botanical intrigue from Perth to Albany. If one were to pause and reflect during the overland journey, one would still feel the exhaustion of surveyors, the exasperation of homesteaders, the exhilaration of naturalists, and the euphoria of botanical artists who reveled in the unparalleled botanical richness of the South-West of Australia.

****

Using this bio-geographic movement and place-based approach as a starting point, and with a focus on the flora of the South-West WA region, *Posthuman Plants* addresses the question of the plant through a number of conceptual prisms, including posthumanist, multispecies, ecocritical, and ecocultural theory. N. Katherine Hayles (1999, 3) argues that the posthuman subject is "an amalgam, a collection of heterogeneous components, a material-informational entity whose boundaries undergo continuous construction and reconstruction." Posthumanist thought emphasizes our co-constituted world as one in which delineations are porous, compromised, and contingent. This volume asks, to what extent does this reconfiguration of human "being" as inherently permeable affect our perceptions of and relationships to plants – those "other" beings that have historically been regarded as part of the scenery and constructed as the foil of animality?

Rather than confined to posthumanist theory, *Posthuman Plants* is situated in the emerging field of interdisciplinary plant studies. The book brings together the dynamic – yet dispersed – body of scholarship referred to variously as "human-plant geographies," "philosophical botany," "human-plant studies," and "critical plant studies." In their shared critique of long-standing human perceptions of flora, these fields attempt to counter and deconstruct the idea that plants can be defined by what they are thought to lack: autonomy, agency, consciousness, and, arguably, intelligence. What unites this broadly ranging text is the assertion that the re-imagining of the vegetal world necessitates the re-conceptualization of plants as active agents in biocultural environments (see, in particular, chapter 6).

The title of the book, "the posthuman plant," resonates in multiple ways. On a more apparent level, "the posthuman plant" is what weaves together the diverse material of this title – an interest in the vegetal that goes beyond single disciplines and specialist discourses, and one that not only provokes but intrinsically requires permeability between the arts, humanities, social sciences, biological sciences, and digital technologies. On another level, "the posthuman plant" connects creative and historical activities to the material being of the vegetal. Our interconnectness with plants reflects our symbiosis with oxygen-producing beings that is at once biological, cultural, social, and linguistic. The book contributes to the increasing dialogue about our understandings of plants and their influence on what it means to be human, more-than-human, and other-than-human.

The text focuses explicitly on "indigenous" plants – defined, in terms of the Western Australian content, as those present in the environment at the time of British settlement in 1829. "Indigenous," "native," or "local" plants (*e.g.* kangaroo paws) are usually contrasted to "exotic," "invasive," or "naturalized" plants (*e.g.* cape daisies) that arrived after colonial settlement, from elsewhere in Australia or from South Africa, Europe, or North America in particular. Exotic plants were introduced to the Perth, WA area unintentionally by colonists (*e.g.* in ship ballasts or cargo) or intentionally by agriculturalists (*e.g.* as food for livestock or to control land erosion). Moreover, some chapters refer to "wild" plants, particularly through Henry David Thoreau's conceptualization of the term (chapter 12). For Thoreau, "wild" is distinct from the Australian notion of "indigenous." Nevertheless, it is vital to recognize that the categories "indigenous" and "wild" are indeed porous, yet also constitute a fixed ontology of plants for many of the theorists, writers, interviewees, and conservationists in this volume.

Posthuman Plants is divided into five sections. Although some of the content of the book is explicitly focused on the vegetal life of the South-West of Australia (chapters 1, 3, 4, 7, and 8) where the author resides, other countries, bioregions, and places indeed figure into the discussion (chapters 2, 5, 6, 9, 10, 11, and 12). Hence, the text aims for regional and international relevance. The chapters are presented as essays on diverse subjects and theories, all organized around the common strand of rethinking the vegetal in contemporary society. The text presents data from ethnographic, auto-ethnographic, historical, and literary sources and draws on theoretical precepts that cross disciplinary demarcations. In previous studies of plants, the transdisciplinary approach to data collection and analysis has been termed "cultural botany" (Ryan 2012). Section I, "Affect and Reciprocity," examines the mourning of the loss of flora in a posthuman world, as well as the current and future value of reciprocity in our interactions with medicinal plant life. Section II, "Heritage and Digitality," provides an interpretation of key concepts of heritage, poiesis, the archive, and digital storytelling in terms of South-West Australian species.

Section III, "Art and Vegetality," provides a critical analysis of the use of live plants in works of digital art and the implications of these productions for notions of plant agency, volition, subjectivity, and intelligence. Section IV, "Poetry and Vegetality," begins with a historical and poetic account of the WA Christmas tree, followed by critical commentaries on Australian and Buddhist ecopoetics. Section V, "Plants and Senses," weighs the possibility of plant-voice and the importance of diverse sensorial modes to our conceptualization and experience of the vegetal world. The Epilogue reflects on nature writing as a vital practice that considers plants the subjects of posthumanist poetic creation.

Part I

Affect & Reciprocity

Johnson Papyrus (5th Century AD). Fragment of an illustrated herbal.

CHAPTER 1

Where Have All the Boronia Gone? A Posthumanist Model of Environmental Mourning

INTRODUCTION: THE VERY BREATH OF SPRING

Loss and hopelessness. This is going to go on and on. We're never going to see these wildflowers again. What I experienced out there is gone. There's still some left, but it's under pressure all the time. All the time. It's sad for me to see this happen. (Kim Fletcher, Perth, Western Australia, April 2013)

Scented or brown boronia (*Boronia megastigma*) is a slender shrub endemic to the South-West corner of Western Australia (WA). Said to possess a "heady, sentimental perfume" (see, for example, Parker 1962, 4), the fragrant blossom is found in the heath lands and eucalypt forests between Busselton and Albany, south of the capital city Perth. Bearing small brown and yellow flowers toward the end of winter (late July–September in WA), boronia was collected in the wild, shipped by train, and sold as an ornamental by Perth street sellers in the early to mid-1900s. In 1947, novelist and columnist James Pollard (1900–1971) wrote of boronia in the Perth newspaper *The West Australian*. Evoking his experience of the wildflower in sensuous terms, Pollard (1947) extols boronia's "perfume stirring memories" (4). He shares a "scented memory" with an onlooker in the street as Pollard – then in his middle years – recollects picking boronia in his youth. In Pollard's account, the flower's aesthetic appeal and the regional economic network, to which it was integral, bridged the divide between city and country: "The young people of the boronia country go out in the rainy dawns of July and August, to gather and dispatch to the city the flower that in this season gives to Perth one its few town cries" (Pollard 1947, 4).

Boronia's "sentimental" olfactory signature sustained cultural memory and identity in urban Perth and in the South-West region for, as Parker (1962) commented, "brown boronia is a West Australian institution – the very breath of spring" (4). With boronia flowers under increasing demand, regulations were implemented during the mid-1900s to control the harvesting of wild plants and the production of fragrant distillations known as "otto of boronia" (Western Mail

1936). However, by the 1960s, concerns about the decimation of boronia began to intensify in popular articles attributing the decline of the species to overharvesting, habitat destruction, bushfires, and the replacement of wild collecting (and the city-country ecocultural networks it fostered) with the promotion of cultivated varieties (Parker 1962, 4). During this time, the viability of the species in the wild became jeopardized. While boronia populations now appear stable (Paczkowska 2013), there are, nevertheless, many cultural losses linked to the decline of the species – as a non-human presence embedded in a network of human-non-human relations. The trains no longer effuse boronia perfume. The street sellers no longer intone the call: "Fresh, sweet-scented boronia! Sixpence a bunch, bor-o-nee-a!"(Parker 1962, 4). The blossoming of boronia – heralding seasonal transition – no longer signifies the rising "breath of spring."

As a result, the cleft between urban and rural life in Western Australia – once sensuously traversed by the "sentimental" perfume of boronia wafted along rail lines – widens with loss. The rapid urban development of Perth now leads to encounters with ersatz boronia where, for example, an online search for the plant genus results in an odd panoply of hits: a Department of Corrective Services facility, an online clothing outlet, an investment company, and a web content management system. Once a "perfume stirring memories," boronia sadly has become a figment of Perth's urban imagination – an artifact confined to the recollections of a generation for whom the wildflower was an "institution" in itself.

PERTH, SOUTH-WEST AUSTRALIA: A CASE STUDY IN BOTANICAL LOSS

> Anstey-Keane Damplands was totally chained by the City of Armadale, which wanted to put a big recreational park there about 40 years ago. Fortunately it was only chained and left undeveloped, so the bush recovered quite quickly after being cleared. But that's why you don't see many old trees. They're all re-growth. Believe me, there were lots of big trees. They won't be seen again in our lifetime because you need 500 years to grow big zamia palms or big banksias. (David James, Forrestdale, Western Australia, September 2009)

With pangs of melancholic nostalgia, I use boronia as a synecdoche and entry point into the broader investigation of the continuing loss of flora in South-

Western Australia, the focus of this chapter. The South-West region refers to the lower corner of the state of Western Australia from Shark Bay in the upper northwest to Israelite Bay east of Esperance in the southeast, and including the metropolitan Perth area. The biodiversity of the South-West – particularly its flora – is of international renown; the region is one of the most botanically diverse Mediterranean ecosystems in the world. A global biodiversity epicenter, the South-West exhibits remarkable biological endemism (Hopper 1998, 2004) and is the only internationally recognized Australian "hotspot" (Conservation International 2008, 2013). Approximately 35 percent of the region's plants are endemic; that is, found to occur in uncultivated conditions only within its bio-geographical boundaries. In the late 1800s, German botanist Baron von Müeller applied the term "botanical province" to the region, observing its unique floristic communities and unusually high rates of endemism (Beard 1979, 107–121).

As the primary urban area within the South-West region, Perth is also a biodiverse, yet rapidly-changing, city. The prominent Australian botanist Stephen Hopper affirms the "tremendous diversity" exhibited by the South-West, distinguishing Perth as "one of the world's most biodiverse cities, especially in relation to plants" (Perth Biodiversity Project n.d., 1). However, the increasing loss of remnant bushland threatens Perth's botanical conservation and compromises the long-term viability of its flora. Indeed, the same factors threatening Perth's unique flora also threaten the plant life of urban areas globally (Farr 2012, Hostetler 2012, Knapp 2010). More specifically, in Australian cities, plant diversity is impacted by a variety of environmental and anthropogenic pressures, including but not limited to bushland development (McKinney 2005), plant diseases (Shearer et al. 2007), dry land salinity (Albrecht 2005, 53, Beresford et al. 2001), and demographic shifts typical of many urban areas (Girardet 2008). Considered one of the world's most ecologically diverse cities, Perth faces the imminent possibility of losing its ecocultural uniqueness, especially in the swiftly expanding southern suburbs between Armadale and Fremantle (Giblett 2006, Giblett and James 2009). While adversely affecting the functionality of ecosystems, these pressures also diminish meaningful human exposure to the natural world and impede intimate interaction – or what Donna Haraway (2008) refers to as co-constitutive entanglement or "becoming with" – between humans and non-humans (4, 88).

Whereas conservation biology and other scientific disciplines address the loss of biodiversity from specialist and technical perspectives, human experience – memory, emotion, sensoriality – bears discernible absences that otherwise go unrecognized and under-researched (Ryan 2012a, Chapter 1). How can we mourn the increasing loss of plant species and their habitats in a quickly transforming urban environment like Perth? Which philosophical and conceptual orientations can guide, nurture, and sustain our processes of mourning environmental loss – on both an individual and collective basis?

This chapter outlines a model of environmental mourning based in connectivity and interdependence – qualities intrinsic to the science of ecology that convinces us, with ever-increasing certitude, of the relational lives of organisms (Smith and Smith 2012). Through a case study of Perth's flora and its human advocates, the chapter propounds a posthumanist theory of environmental mourning – one that relies on notable developments in multispecies theory and the ecological humanities. A posthumanist theory extends recent work, conducted from different disciplinary perspectives, by Glenn Albrecht (2005, 2010, Albrecht et al. 2007), Donna Haraway (2008), Timothy Morton (2010), Deborah Bird Rose (2008, 2011a, b, Rose and Robin 2004), Thom van Dooren (2010, 2011), Cary Wolfe (2010), and Kathryn Yusoff (2011), as well as previous research into environmental mourning and the cultural significance of plants in the South-West region (Ryan 2012a, chapter 11, Trigger and Mulcock 2005).

Forwarding a non-binary position between the human, "unhuman, nonhuman, or 'more-than-human', or possibly even inhuman" (Morton 2010, 252), a posthumanist model of environmental mourning will be developed in respect to the loss of botanical habitats in Perth and the impact of this loss on the sense of place of interview respondents. The ethnographic material that informs this chapter represents a small part of a series of interviews performed since 2009 with South-West Australians in order to understand their commitments to botanical conservation in the context of Perth's declining flora. What started as a one-dimensional interest in the interviewees' aesthetic attitudes toward South-West plants – their beauty, sublimity, picturesqueness, or ugliness and the relationship between environmental aesthetics and community appreciation of the region's biodiversity – evolved into a multi-nodal study of human sensoriality and flora. During the course of previous research, memory and mourning have been implicit formations, generated at the intersection of scientific knowledge, sensory experience, and emotional recollection. Throughout this process the researcher found it impossible to disentangle memory and mourning from the understanding of how one's sense of place develops in relation to scientific and non-scientific understandings of plants – a broader phenomenon termed "floratopaesthesia" or literally the sense of place fostered through sensory experience of flora (Ryan 2012a, chapter 13).

The Humanist Aspects of Freudian Mourning

> Some of the orchids, like the spider orchids, are not as plentiful, especially the white spider orchids. Weeds smother small plants. Dodder laurel creeper climbs on paperbarks. It's still here. Many local swamps have been bulldozed in the last thirty years. Weeds have changed the appearance of the lake forever. I don't know what the answer is. (David James quoted in Giblett 2006, 97)

A multispecies theory of environmental mourning eschews the problematically individuated and human ego-focused principles of Freudian mourning (Freud 1968). A theory of environmental mourning should transcend the individuated subjectivity of Freudian mourning that positions an object of loss – a habitat, a plant, an animal – in a dichotomizing relationship to the grieving ego of the subject – a human mourner, a conservationist, an interviewee. This section begins by critiquing the insufficiency of Freudian mourning to account for the loss of the networks – interdependencies, connectivities, relationships – between living creatures and their living and non-living milieux. Freudian mourning foregrounds human subjectivity and, consequently, backgrounds the ecocultural interconnections and multiple subjectivities with which a human subject perpetually transacts. Moreover, Freudian mourning posits an exceedingly anthropocentric mode of mourning – a critique pursued in relation to the work of psychoanalytic scholars (Clewell 2002, Dodds 2011, Homans 1989). In contrast, a theory of environmental mourning should account for the loss of the nodes themselves and non-human "things" with which a human subject has established ecological intimacy – in ways that inherently counter tendencies toward human narcissism throughout the different manifestations of environmental mourning.

Freudian mourning involves a centering of the human subject in tandem with a monologic notion of normative subjectivity. As these tenets are also shared by humanism, this section will briefly explore the interstices between humanism and Freudian psychoanalysis, with which some scholars have associated Enlightenment ideals (for example, Dodds 2011). Humanism refers broadly to an aggregate of philosophies that prioritize human agency, empiricism, and rationalism over fideism or faith-based principles of meaning-making. As Fowler (1999) points out, "humanism is concerned with the secular, rather than the religious; with this life, as opposed to projections of life beyond death when the *human* no longer exists; and with the immediacy of the temporality of human existence rather than any suggestions of the eternal nature of the human [emphasis in original]" (9). For Fowler (1999), humanism reached its apotheosis in the 17th and 18th century Enlightenment ideals of Kant and Newton (16). Hence, humanism emphasizes "empirical science and critical reason" (Wolfe 2010, xiv), over the supernatural or divine, in learning about and interacting with the natural world. In his critique of humanism, Cary Wolfe (2010) contends that the notion of "the human" is constructed in opposition to animality in order to liberate the human from biological and evolutionary realities – and to transcend the immanence of embodiment (xv). Humanism pivots on "normative subjectivity – a specific concept of the human" (Wolfe 2010, xvii), which polarizes the human and the animal and, consequently, hyper-separates humanness from vegetality.

While an extensive treatment of Freudian psychoanalysis – including Freud's other writings on grief in "On Transience" (1916) and *The Ego and the Id* (1923) – is out of the scope of this chapter, some of the humanistic tendencies of Freud's theory of mourning in the essay "Mourning and Melancholia" (1917) will be identified. The critical position developed here is that Freud's distinction between "conscious" mourning and "unconscious" melancholia provides a conflicted basis for a theory of environmental mourning accounting for connectivity and interdependence between species – one which circumvents the pitfalls of humanistic subjectivity and egocentrism that multispecies theory and the ecological humanities seek to redress. Early in "Mourning and Melancholia," Freud (1968) defines his mourning framework as comprising both tangible and abstract "objects" or as "regularly the reaction to the loss of a loved person, or to the loss of some abstraction which has taken the place of one, such as one's country, liberty, an ideal, and so on" (243). Whereas mourning is a normal grieving process in response to loss, which when "completed the ego becomes free and uninhibited again" (Freud 1968, 245), melancholia is a pathological outcome (Clewell 2002, 43). As an unconscious response to loss, melancholia entails "profoundly painful dejection, cessation of interest in the outside world, loss of the capacity to love, inhibition of all activity, and a lowering of the self-regarding feelings" (Freud 1968, 244).

Freud's theory of mourning pivots on a binary between mourning as the extension of grief at the loss of the world and melancholia as an inward projection of grieving emotions linked to human narcissism: "In mourning it is the world which has become poor and empty; in melancholia it is the ego itself" (1968, 246). For Freud, the process of mourning oscillates between the mourning subject and the mourned object, whereas melancholia inverts grieving emotions into a state of solipsistic self-reflection in which the world and human-non-human interconnectedness vanish. Cultural studies scholar Rod Giblett (1996) rephrases the Freudian distinction as "in mourning the world is experienced as loss whereas in melancholia the ego is experienced as loss" (177). Significantly, Freudian mourning posits the environment as an object of a human subject's mourning – a backdrop to ego-driven loss that, when pathological, instigates the experience of melancholia. Freud articulates the human grief process through which the mourner severs emotional ties to the lost object, entailing a "detachment of libido" (Clewell 2002, 44) marking the completion of mourning. In the final analysis, mourning ceases when the mourner dissolves emotional linkages to the lost object-cathexis. The severance liberates the human ego to focus on a new object, creating a substitution for the absence of the lost one – a process that seems to be driven by narcissistic self-identification rather than an ecological sense of interdependence.

The dyad of Freudian mourning and melancholia circumscribes the traditionally unified subject in which human allegiance is to personal desire rather than to the "objects" of desire (Clewell 2002, 1) and even less so to the communities or networks of desire, affection, and love (Rose 2008, 2011a). For the mourning of plants, what is lacking in Freudian thought is an expression of human interrelationship to the world beyond the pathologies of the strongly individuated subject entrapped in a possibly self-perpetuating cycle of melancholia. Clewell (2002) argues that "Freud's mourning theory has been criticized for assuming a model of subjectivity based on a strongly bounded form of individuation" (1). In the context of botanical loss, the threatened, endangered, or extinct plant is mourned, not out of the intrinsic loss of the plant and its ecocultural embeddedness, but rather as a reflection of the subject's ego-focused inevitable decline. Strikingly anthropocentric, Freud's "bounded form of individuation" reflects a discourse that requires object-subject dynamics to produce transferences between the mourning subject and the mourned object. As such, Freudian mourning limits the development of a form of environmental mourning based in qualities of connectivity and interdependence.

In comparable terms, environmental psychologist Joseph Dodds (2011) asserts that Freudian psychoanalysis reflects the "Weltangschung [worldview] of science and the Enlightenment" (31). As Dodds argues, Freud's writings on mourning exhibit the predominant binaries of his era, which position the imperatives of psychoanalysis in opposition to the drives of the natural world: "The principle task of civilization, its actual raison d'être, is to defend us against nature. We all know that in many ways civilization does this fairly well already, and clearly as time goes on it will do it much better" (Freud quoted in Dodds 2011, 31). As Dodds goes on to explain, the Freudian dichotomies – between masculinity and femininity, civilization and nature, order and chaos, and autonomy and dependency – are definitive tropes found in the thinking of the Enlightenment, modernity, and patriarchy. Freudian psychoanalysis, therefore, carries elements of the Enlightenment project to bring unconsciousness to consciousness, lending greater weight to the rational ego and resulting in a "master-slave system of absolute binaries, and an attempt to maintain an illusory autonomy and control in the face of chaos" (Dodds 2011, 32).

As a counterweight to Freudian mourning and its humanistic inclinations, "multi-species mourning" will be theorized next in relation to the loss of Perth's flora through theorists associated with three domains: posthumanism, the ecological humanities, and multispecies theory. While examining how concepts of environmental mourning have hitherto been developed by environmental scholars, independently of humanistic traditions, the section will argue for the applicability of these three paradigms – located in literature, philosophy, cultural studies, and animal studies – to an emergent perspective on flora called "human-plant studies" (Ryan 2013, chapter 6), which most notably integrates emerging notions of plant subjectivity into critical humanities-based botanical research (Marder 2013a).

TOWARD A POSTHUMANIST MODEL OF ENVIRONMENTAL MOURNING

> Wreath flowers (*Leschenaultia macrantha*) were rare. If you picked those, by the time you stood up, it started to wilt. An amazing flower – just pick it up and it's gone. They were up around Mullewa and Bullardoo Station. There's a big hill there. It was iron ore and, sadly, they loaded the hill onto trucks and railway cars and carted it away. Tallering Peak was a really important place. (Noel Nannup, Mount Lawley, Western Australia, July 2010)

Rather than Freudian subject-centered mourning, a posthumanist framework relates the mourning of declining plants to bodily experience and sense of place while counterbalancing anthropocentrism through the decentering of human subjectivity. Indeed, as one of its major aims, posthumanism negotiates the "decentering of the human" (Wolfe 2010, xv) in the construction of knowledge and meaning. Recent posthumanist developments in multispecies theory – most prominently the notion of "companion species" (Haraway 2008) – and in the ecological humanities – specifically the concept of "connectivity ontology" (Rose and Robin 2004) – offer posthumanist perspectives for theorizing the mourning of vanishing non-humans. Moreover, in decentering human subjectivity without nullifying the essential role of experience, a productive direction proffered by multispecies theory is a focus on human-non-human sensory entanglements. Barad (2010) stresses, "entanglements are not a name for the interconnectedness of all being as one, but rather specific material relations of the ongoing differentiating of the world. Entanglements are relations of obligation – being bound to the other – enfolded traces of othering" (265). This section departs slightly from Barad in her association of entanglement with material obligations only. Human-non-human entanglement also entails intangible mnemonic, emotional, and sensual interconnections that develop in conjunction with ongoing material transactions (see chapters 3 and 4). In this context, posthumanist mourning is inclusive enough to account for both human-non-human entanglements and the plant species impacted by extinction pressures, such as habitat loss, as well as the often devastating emotional and phenomenological depletion of human sense of place that occurs.

In order to articulate more clearly a theory of environmental mourning, this section will outline some of the tenets of posthumanism in order to clarify its relevance to the decline of a biodiverse world. Through the decentering of subjectivity, posthumanist environmental mourning negotiates the positioning of the human in the sensuous fabric of the world, in the materiality of an interdependent ecological existence. For Wolfe (2010), posthumanism "opposes the fantasies of disembodiment and autonomy, inherited from humanism itself" (xv). In terms of Barad's notion of entanglement, posthumanism embeds human subjectivity in the material demands of place through embodied experience and awareness of tacit human-non-human transactions and interconnections. While the term "posthumanism" emerged in the mid-1990s in the humanities and social sciences, a notable antecedent is systems theory, developed by Gregory Bateson, Norbert Wiener, and others during the 1950s (Wolfe 2010, xii). Early developments in system theory attempted to dislodge human subjectivity "from any particularly privileged position in relation to matters of meaning, information, and cognition" (Wolfe 2010, xii). But where does environmental mourning lie within Wolfe's interpretation of posthumanism, which is largely contextualized through human-animal studies? The first motion toward a posthumanist model of botanical mourning, therefore, is to interface Wolfe's analysis of the decentering of the human with philosophical (Marder 2013a), literary (Ryan 2012a, 2013), and scientific (Trewavas 2002, 2003, 2006) developments in plant research, in order to reach a theory of botanical mourning applied to interpret some of the interviewees' comments in the following section.

In addition to its focus on relationality, posthumanist mourning is specific (multispecies or interspecies); the mourning of botanical loss tends to involve intimate knowledge of actual species. Acknowledging the layers of ecocultural loss embedded within human-non-human entanglements, posthumanist mourning provides scope for the mourning of individual non-humans in order "not to reduce them to interchangeable cogs in an ecosystem machine" (Van Dooren 2010, 273). Indeed, as the following section will demonstrate, the contraction of plant diversity underlies the impoverishment of interviewees' sensory lifeworlds, memory networks, and emotional wellbeing. While posthumanist mourning will always remain human – after all, we are human beings who mourn and whose affective relations contract when other species are lost – it can engender within us an awareness of Freudian mourning's dangerous hyper-subjectivity. The decentering of the human is not simply a turn from anthropocentrism to ecocentrism (from culture to nature), but rather a return to a sense of relationality between species as the essence of mourning. This is not to say that humans are the only beings who mourn: other mammals do and, if considered to have emotions, plants and fungi as well. Although important to take into account and explore further as part of a posthumanist theory, the mourning experienced by plants is, unfortunately, out of the scope of this chapter. The potential of posthumanist environmental mourning, therefore, is its capacity to reflect the "mosaic quality that recognizes causality between the self, the other and the environment" (Ryan 2012a, 278) – its appreciation that environmental loss fragments multispecies entanglements and renders human experience of place monochrome and life-less (literally). The crucial interspecies dimension of environmental mourning, furthermore, encompasses the roles of exotic or invasive flora – exemplified by the intrusion of yangee rushes into endangered Forrestdale Lake wetland ecosystems south of Perth (Giblett 2006). Interspecies mourning comes to reflect the dynamics of endangered places, including, in the case of Forrestdale, the colonization of a wetland environment by an aggressive plant, the ensuing displacement of indigenous biota, and the decline of the aquatic ecosystem over time. In many examples of environmental loss, the human or non-human "aggressor" (i.e. government agency, corporate developer, introduced animal or exotic weed), such as the yangee rush, becomes an object of anger and despair as much as the lost species becomes an object of mourning and grief. A posthumanist model suggests that the understanding and experience of mourning can become intrinsically ecological, moving from the disappearance of individual species or the impoverishment of the human life-world towards an ecology of mourning that is much more complex, inclusive, intersubjective, evocative of place, and deserving of further research.

In terms of its specificity, the title of this chapter, "Where Have All the Boronia Gone?", responds in part to Timothy Morton's work on elegy and his development of "dark ecology" with respect to environmental loss. In a section "Eco-elegy, or Where (Will) Have All the Flowers Gone?" Morton (2010) propounds an essentially negative view of environmental mourning, claiming that "we cannot mourn for the environment because we are so deeply attached to it – we *are* it [emphasis in original]" (253). In other terms, Morton (2010) contends – albeit facetiously and in response to Judith Butler's analysis of melancholia – that "the truest ecological human is a melancholy dualist, mourning for something we never lost because we never had it, because we *are* it [emphasis in original]" (253). Through statements such as these, Morton firstly deconstructs "environment" as a signifier, secondly suggests the impossibility of environmental mourning, and thirdly attempts to counter the tendency in nature writing to construct the natural world as a backdrop for human grief and – tangentially – as an object-cathexis of human mourning in Freudian terms. Importantly, Morton (2010) asks, "what happens when this backdrop becomes the foreground?" (253). Providing no firm alternative of his own, Morton characterizes environmental mourning in its current state (*read* Freudian) as a "narcissistic panic [that] fails fully to account for the actual loss of actually existing species and environments" (Morton 2010, 255). Although deconstructive and ultimately nihilistic, Morton's critique does invigorate the need for a theory of environmental mourning drawn from multispecies thinking – one which goes beyond the generalizability of the term "environment" and alternately accounts for "actually existing species" in all their specific materialities and sensuous modes. Moreover, although eliding crucial distinctions between human and non-human subjectivities – "we are it" – Morton lodges implicitly a critique of Freudian mourning and, hence, calls for a new theory of environmental mourning, one reflecting the philosophy of ecology as "thinking how all beings are interconnected, in as deep a way as possible" (2010, 255).

The posthumanist approach proposed here also looks toward constructive, community-based, and human-focused interpretations of environmental mourning, exemplified by Glenn Albrecht's (2005) research into solastalgia or "the pain or sickness caused by the loss or lack of solace and the sense of isolation connected to the present state of one's home and territory" (48). Whereas nostalgia refers to one's sense of loss when displaced, solastalgia encompasses the feelings of mourning when one experiences firsthand the deterioration of his or her home place or bioregion and, as Albrecht (2016) further explains, reflects the "deep form of existential distress when directly confronted by unwelcome change in [a] loved home environment" (216). Solastalgia is the "lived experience of the loss of the present" (Albrecht 2005, 48) linked to the decline of the capacity of a place to nurture a sense of home or a feeling of love as "topophilia" (Tuan 1974) or "biophilia" (Wilson 1986). Elsewhere, Albrecht et al. (2007, S96) define solastalgia as "place-based distress in the face of the lived experience of profound environmental change." Evolving from ethnographic work in the Upper Hunter Region of New South Wales, Australia, solastalgia is the "homesickness one gets when one is still at 'home'" (Albrecht 2005, 48) and is most intense as a result of sustained sensory and emotional experience of the environment's decline over time:

> The people of concern are still 'at home', but experience a 'homesickness' similar to that caused by nostalgia. What these people lack is solace or comfort derived from their present relationship to 'home', and so, a new form of psychoterratic illness needs to be defined [...] solastalgia exists when there is the lived experience of the physical desolation of home. (Albrecht et al. 2007, S96)

Indeed, the factors precipitating solastalgia can be either environmental or anthropogenic, and can include "drought, fire and flood" as well as "land clearing, mining, rapid institutional change" (Albrecht 2005, 48), the latter most evident in the following interviews with Perth conservationists.

While Albrecht's solastalgia accounts for the "psychoterratic" effects of generalized place-based deterioration, however, posthumanist mourning provides a conceptual perspective for addressing the decline of "actually existing species" (Morton 2010, 255) – plants in this instance. Environmental mourning accounts for the emotional bonds between people and places in response to what Porteous(1989)terms "topocide" (230) or the annihilation of place and what Giblett (1996) calls "aquaterracide" (68) or the killing of wetlands. The specific focus of posthumanist environmental mourning – on urban botanical loss here – foregrounds the emotional and material connections people develop to flora. Additionally, while Albrecht's solastalgia has been developed in reference to the effects of mining and drought, the loss of Perth's flora expands the contexts of despair in which human wellbeing is impacted adversely by environmental loss. As proposed in this discussion, a posthumanist approach to environmental mourning examines the nature of loss itself. Yusoff (2011) argues that "loss can be considered a measure of feeling that proceeds from vulnerability, from the loss of a self that is constituted in relation, rather than in isolation" (579). Following Yusoff, multispecies mourning brings attention to the constitution of human-non-human subjectivities and the relational quality of loss beyond the anthropocentric Freudian dyad of mourning and melancholia.

This model of environmental mourning also follows recent work by ecological humanities scholars on the social and philosophical impacts of extinction. In his research into the rapid decline of Indian vultures, Thom van Dooren (2010) summarizes poignantly the significance of a multispecies perspective: "Attentiveness to the relationality and interdependence of life is particularly important because the death, and subsequent absence of a whole species, unmakes these relationships on which life depends, often amplifying suffering and death for a whole host of others" (273). Significantly, responses to place-based deterioration often involve the scientific management of populations and species, rather than an acute awareness of the pain experienced by individual humans and non-humans. Van Dooren (2010) insists that "individuals – as ethical subjects – be brought back into a conservation discourse that is saturated with species, habitats and ecosystems" (277). A posthumanist model of environmental mourning acknowledges individuals as "relational beings" and that mourning, as a consequence, must be relational, multispecies, and multi-sensorial in order to be emplaced as a phenomenon in itself in the world.

MOURNING THE DECLINE OF FLORA IN SOUTHERN PERTH: VOICES FROM THE COMMUNITY

> What we took for granted is now threatened by housing and the expansion of agriculture and roads. The piece of bush that we took for granted as kids is now being threatened. It's a bitter shame. I spend more time trying to protect wildflowers than actually going out trying to enjoy them for what they are. (David James, Forrestdale, Western Australia, September 2009)

This section will further advance a posthumanist model of environmental mourning through analysis of interviews with residents of the southern Perth suburbs. Since 2008, interviews have been conducted with botanists, conservationists, educators, wildflower enthusiasts, and others with long-standing relationships to the city's rapidly disappearing plant life. Drawing on ethnographic methods and interrogating theories of human memory, I have found that the decline of plant species comes with the loss of olfactory, gustatory, haptic, acoustic, and visual networks between humans and non-humans (Ryan 2012a, chapter 8). This discussion will focus in detail on two interviews with Armadale area conservationists David James and Kim Fletcher – both of whom express the complexities of environmental mourning, yet from differing perspectives. The interviewees are well-informed, passionate about, and closely involved with Perth's flora; for example, Kim Fletcher has collaborated with Western Australian scientists in the taxonomic identification of orchids. In addition to the interviews, previously published oral histories will be mentioned (Giblett 2006) and, where relevant, archival and literary material, exemplified by James Pollard's boronia writings. The eclecticism of this chapter as a whole reflects an integrated approach to humanities-based plant research, examining a wide range of sources and theorized as "cultural botany" (Ryan 2012a, chapter 1).

Themes of loss are evident in an interview with David James of Forrestdale, Western Australia. Born in the early 1950s, a passionate activist and self-trained botanist, David has lived near Forrestdale Lake all his life. Forrestdale Lake is located on the Swan Coastal Plain, shared by Perth to the north, and protects numerous indigenous floral, faunal, fungal, and insect communities (Giblett 2006). Nearby Anstey-Keane Damplands is one of the most botanically significant places on the Swan Coastal Plain and more biodiverse than popular Kings Park, adjacent to Perth Central Business District (Giblett and James 2009). Anstey-Keane lies at the northern tip of the Pinjarra Plains, a system of flat damplands – moist, shallow sinks – including the most desirable soil on the Swan Coastal Plain for pasture and development (Beard 1979, 27). A member of several local conservation organizations, David conveys his exasperation over the ostensibly insurmountable pressures on the environment posed by suburban development. For David, Forrestdale Lake Nature Reserve and Anstey-Keane Damplands constitute "emotional geographies" (Bondi, Davidson, and Smith 2005) – places linked to his extensive conservation commitments in the quickly changing southern areas of Perth. David's involvement with different organizations charged with protecting land is "a period of, shall we say, activism [with] organizations that are actually trying to preserve the environment." Environmental activists, such as David, exemplify a single-minded devotion to maintaining a community-based sense of place through the protection of plants.

During David's childhood in the 1950s, the bush around Perth seemed limitless and impervious to modern suburban expansion: "We'd walk through bush to catch the school bus." His memories suggest a popular perception of the flora as all-encompassing and inexhaustible at that time. The former abundance of the bush intermingles with recollections of the encroachment of development and his attendant emotions of powerlessness:

> In those days, the bush was everywhere. Nowadays, you realize how threatened it is, but in those days, it was common. We'd walk through bush to catch the school bus. We took it all for granted. Even in those days, people were destroying bushland but, because there was so much, as a kid, you just accepted it.

David's recollections reveal how environmental mourning – in Albrecht's (2016, Albrecht et al. 2007) terms "solastalgia" or the distress of witnessing the decline of one's home place while one is still there – can involve detailed information about the land transformed by the juggernaut of development. His childhood reverie has been displaced by an anxiety over the manifold threats to the local landscape, compelling him to align with conservation initiatives: "I spend more time trying to protect wildflowers than actually going out trying to enjoy them for what they are."

David's interview also demonstrates environmental mourning as a wellspring of emotions about the natural world. Moreover, the emotion-filled memory narratives of amateur botanists and local conservationists like David can impart a sense for the scale of change. For example, the proliferation of exotic plant species on roadsides is a distinctly noticeable change over David's lifetime: "Years ago roads were narrow and roads with vegetation in good condition were quite normal." His recollections confirm the progressive incursion of exotic plants on the west side of Forrestdale Lake and the gradual disappearance of certain orchid species. These characteristic intrusions to the composition of his home place have occurred within his memory:

> Nowadays this side of the lake's pretty much weed infested, but in those days the weeds weren't quite so bad. We used to get orchids growing alongside the road here, spider orchids and different species growing amongst the weeds.

Slow-growing plants, such as banksias and the zamia palms, have been severely affected by destruction of bushland areas, specifically Anstey-Keane Damplands and Forrestdale Lake (Giblett 2006). According to his direct experience, the climax character of the land has been permanently altered, despite positive claims about the merits of bushland reconstruction by environmental restoration techniques:

> After 30 or 40 years, it looks quite natural. But believe me, if you went
> back before that, there was a lot of big stuff in there. That won't be seen
> again in our lifetime because you need 500 years to grow big zamia
> palms or big banksias.

David evokes one of the distinguishing qualities of the South-West flora: its
ancientness and slow-growing habits resulting from low soil nutrients and
intensive solar exposure (for example, Breeden and Breeden 2010). Some plants
require hundreds of years to reach a mature state, amplifying environmental loss
in a manner evocative of the mourning of old-growth forests farther south in the
South-West region (Trigger and Mulcock 2005).

Like David James, Kim Fletcher has lived most of his life near Forrestdale
Lake Nature Reserve and Anstey-Keane Damplands near Armadale in the
southern suburbs of Perth. Kim was born in 1937 in Cottesloe, Western Australia,
to a family of wildflower enthusiasts and amateur orchidologists. As an educator,
in Armadale and elsewhere in the state, he developed strong pedagogic skills he
now uses in his role as a volunteer guide with Kings Park and Botanic Garden.
Some of Kim's earliest memories of Perth's wildflowers relate to orchid
collecting on Sunday afternoons with his father and mother. In order to locate
wildflowers, Kim's family frequented swathes of original bushland between the
railway line and the main highway to Perth where, in particular, orchids
proliferated. However, as Kim explains, much of this land has now been
converted to suburbia:

> It was ideal for collecting orchids, particularly spider orchids
> [*Arachnorchis* spp. formerly *Caladenia* spp.] and enamel orchids
> [*Elythranthera* spp.], because it was sandy country. This, of course, has
> now disappeared altogether as bushland. It's covered with suburbia, with
> houses now. That started probably in the late 1950s and early 1960s.

Kim evokes a sense of multispecies mourning in the following excerpt in which
he describes locating a rare hammer orchid (*Drakaea* spp.). As an expression of
posthumanist mourning, his recollection involves other species – a marri tree
(*Corymbia calophylla*) and cormorants (*Phalacrocorax varius*) – as well as
sensuous detail about the attractive qualities of the orchid. The land, converted to
suburbia, has been shared amongst different species as a community;
consequently Kim's expression of mourning exhibits relational, multispecies, and
sensory aspects. However, his initial curiosity and child-like awe shift abruptly to
hopelessness – hopelessness of never finding the orchids in his home place again:

Under a marri tree, I found a heart-shaped, glossy green, beautiful leaf lying flat to the ground with a stalk coming up. I thought, 'Oh, what's this? It must be an orchid'. I visited practically every day and eventually, of course, it came out. It happened to be a hammer orchid, the first I'd ever seen. Apparently now these are declared rare flora. I did find the same orchid elsewhere over in the Forrestdale Lake area, but they were the only two I've ever found. There's no hope of finding them because the land is degraded badly and is going to be surrounded by suburbia in the near future. Incidentally, it was also a rookery for cormorants.

The multispecies ambit of Kim's mourning furthermore includes references to spearwood (*Kunzea ericafolia*) and the intrusion of exotic plants, such as arum lilies (*Zantedeschia aethiopica*), leading to the formation of hybrid botanical communities intermingled with the new housing estates. Kim expresses other feelings of loss – of returning to his home ground in hope of finding jewel orchids but repeatedly being disappointed. In the following excerpt, there is also an obvious connection in Kim's memory between spearwood and jewel orchids (*Anoectochilus* spp.), indicative of the multispecies community he now mourns:

There were *Kunzea ericafolia* growing along Forrest Road. They were so common. Of course, during the 1980s, it was cleared and houses put on it. There is still some bush in the area, but it's been invaded by arum lilies and goodness knows what. That site has gone, I'm sorry to say. Although Armadale Golf Course is contiguous to the conservation area around Forrest Road, the nature reserve and the orchids have disappeared. I've been back many times trying to find jewel orchids.

Local flora, especially orchids, signify place in Kim's lived experience – reflecting the development of floratopaesthesia or sense of place through plants. Kim narrates his experience of finding ghost orchids or rattle beaks (*Lyperanthus* spp.) in Wungong Gorge east of Armadale. He later came across a surprisingly large cluster of ghost orchids in the city of Armadale, not far from Wungong. Along with the ghost orchids, Kim recalls the blue china orchid (*Caladenia gemmata*) and diverse species of banksia in the Peel Estate area outside Armadale. As an orchid lover, the ghost and blue china orchids represented peak, affective, and multispecies experiences for Kim – "it took my breath away." However, these sites of memory and meaning have been razed:

Not as far as the bush to the east of Armadale, which is forest country, I would go exploring. One of the big finds was in Wungong Gorge. I found a patch of *Lyperanthus*. It's called the ghost orchid or rattle beak, but that's long gone. However, in Fletcher Park [in the City of Armadale] I found about 40 ghost orchids. I also remember getting out into the Peel Estate where there were *Caladenia gemmata* under the banksia. It is a little blue orchid and that was right through. It took my breath away when I found that. But it's all been cleared.

In another excerpt, Kim's memories of rabbit or hare orchids (*Leptoceras menziesii*, formerly *Caladenia menziesii*) capture the sensuousness of the species growing together in a mass. Nevertheless, the poignancy of Kim's recollection is tinged with powerlessness and despair, marked also by an awareness of the "weeds" that have displaced the orchids:

There was a huge patch of what was formerly known as *Caladenia menziesii* or the rabbit orchid. It was a big mass. All the leaves and flowers were shimmering right there. Many times I've gone back with a tear in my eye, looking at where they were among the weeds close to the road.

Solastalgia is an apposite framework to describe Kim's mourning of environmental loss. Infused with emotional and family resonances, his home place has changed drastically as the orchids he loves disappear at alarming rates: "It's a very sad time for me when I go back to these places that I used to frequent so often as a kid when I got to love orchids." Kim's statements underscore the idea that environmental mourning is ecological or, in Morton's terms, "interconnected, in as deep a way as possible" (2010, 255). The mourning of a plant species is rarely of that species in isolation but rather within its ecological milieu – within its multitude of relations. For Kim, the orchids of the Perth suburbs invigorate recollections of other plants, animals, and birds, as well as a sense of place and personal identity constituted over many years of living in the Armadale area. As Yusoff aptly states, loss "requires mourning and grieving for the destruction of a relation and those subjects that are constituted through that relation" (2011, 579).

Highlighting the home place shared by Kim and David, Rod Giblett's *Forrestdale: People and Place* (2006) is based on oral histories with Forrestdale Lake residents. Many of the interviewees intimate that the mourning of botanical loss also involves the mourning of the sensory environments in decline as plants become less common. Posthumanist multispecies mourning, therefore, in part responds to human grief caused by the impoverishing of sensory landscapes for, as Wolfe (2010) argues, posthumanism "opposes the fantasies of disembodiment and autonomy" (xv). For example, Katherine Taylor Smith was born in 1905, lived near Forrestdale Lake until 1917, and passed away in 2004. She recounts childhood experiences from the 1910s and observes the decline of kangaroo paws (*Anigozanthos manglesii*) and other plant species in the area:

> They used to grow thick on the sides of the road. It's nothing like that now. Nothing. I mean the fumes from motorcars are what kill everything [...] There were many more wildflowers. Kangaroo paws and waxy plants and purple waxes and spider orchids. There was everything. Donkey orchids were will still growing in Taylor Road not so many years ago. (quoted in Giblett 2006, 88–89)

Conversely, Katherine's sense of mourning indicts exotic species as factors of negative change: "In those days there was white sand around the whole of the lake and none of those yangee rushes [*Typha orientalis*]" (89). Moreover, her emotional affinity for Forrestdale Lake is evident in her vivid sensory recollections: "Oh I loved it. I loved the life we had. It's just different, smells different. The country smell was beautiful. The kangaroo paws were the most outstanding smell but they weren't the most beautiful flower" (91).

The qualities of interdependence and connectivity vis-à-vis botanical mourning, evident in these interviews, are further summarized by Don Williams, proprietor of Hi-Vallee Farm in Badgingarra, WA, north of Perth. Don describes a form of multispecies mourning, in which the loss of one species has broader implications for the whole system of which it is part, including other plants, animals, and insects:

> No one can put a value on individual species. Some people will say if we lose it, it doesn't matter, more will evolve. It would appear that they won't evolve as quickly as we can wipe them out. What a lot of people forget is that if you lose a plant species, you could lose an animal or insect species. And a lot of the orchids have evolved around one individual insect which pollinates them.

Don suggests the importance of environmental mourning as a relational response to habitat decline, in distinction to Freudian mourning in which the object in isolation becomes the focus of emotional transference, narcissistic self-identification, and ultimately object-cathexis severance. In sum, the exploration of memories of plants invariably confronts the mourning of the loss of biodiversity. As conservationists, David James and Kim Fletcher have been life-long residents of the Armadale area, and have developed emotional attachments to plants through direct experience of their home places. They express a startling sense of the once-biodiverse Perth suburbs in a process of transformation at the hands of numerous seemingly insuperable economic and environmental forces.

CONCLUSION: TRANSFORMATION THROUGH LOSS IN A MULTISPECIES WORLD

This chapter has outlined three components of a posthumanist model of environmental mourning: relationality, multispecies entanglement, and sensoriality. As the interviews with David James and Kim Fletcher indicate, posthumanist mourning decenters the human without abnegating his or her emotional, sensory, and mnemonic connectedness. In fact, the posthumanist model developed in this chapter could be said to intensify human expressions of affect toward other beings – as human-non-human interstices are valued and made more palpable through this mode of thinking about mourning (i.e., the ecologically situated human subject within his or her relations). In exploring the interviewees' emotional responses to botanical decline, the model does not fall into the bind of human subjectivity it wishes to transcend. Instead, the impoverishment of the interviewees' lifeworlds signifies the loss of their relational situatedness: in the echo of the landscape's emptiness are the voices of the orchid, boronia, and banksia.

Along similar lines, Mick Smith proposes a posthumanist concept of ecological community. Smith (2013, 21) outlines four "materially inseparable" aspects that can assist us in articulating multispecies loss: "material manifestation (appearances), material involvement (effects), semiotic resonance (meanings) and phenomenological experiences." A loss of a species is, therefore, the loss of a species of appearance; a species of creative involvements; a species of significance and openness on the world; and a species of ecological community (Smith 2013, 22). In other words, the disappearance of a plant species always entails a rupture of community networks, constituted by the presence of different beings, including plants and humans. Posthumanist mourning attempts to think across the categories of "human" and "plant" altogether, treading a middle way between anthropocentrism, on the one hand, and ecocentrism, on the other.

However, a future-oriented question lingers: How can environmental mourning be made into a resource for constructive change and ecological recovery? Although not in terms of biodiversity loss, Judith Butler has argued that mourning can be leveraged as a "resource for politics" (Yusoff 2011, 578). Although still an elusive and intangible process by all accounts, mourning, she suggests, "has to do with agreeing to undergo a transformation (perhaps one should say *submitting* to a transformation) the full result of which one cannot know in advance [italics in original]" (Butler 2004, 21). While this transformation can be personal, it can also be multispecies, just as environmental loss reflects non-binary human-non-human entanglement. A posthumanist model of environmental mourning, therefore, opens pathways for recognizing, managing, and recovering from place-based loss so that positive transformation can be catalyzed in response to the severe alteration of home places and habitats. For example, the restoration and conservation of botanical communities returns a vital heritage of plant-based sounds, smells, tastes, sensations, and sights, thereby promoting greater human-non-human well-being in urban areas (Millenium Ecosystem Assessment 2005) (also see chapters 3 and 4).

The obverse side of environmental loss is that the activism of James and Fletcher is in part catalyzed by their mourning. The values of posthumanist mourning ensure that transformative actions in response to loss are governed by a concern for plants, animals, and insects and their protection, restoration, or memorialization (Ryan 2012a, chapter 11). For example, the Keane Road Strategic Link (KRSL) has been proposed by the Armadale government to bisect Anstey-Keane Damplands, a federally acknowledged threatened ecological community. The road extension would disastrously impact rare and endangered species, such as the critically endangered native short-tongued bee (Neopasiphae simplicior) and the endemic graceful sun-moth (*Synemon gratiosa*). As this example demonstrates, multispecies thinking is integral to environmental activism campaigns, such as those organized by James and the Friends of Forrestdale, and reflects a desire among conservationists to keep intact communities of plants, animals, and insects for the inherent value of those species and the wellbeing of humans. A posthumanist model of mourning is able to articulate these considerations.

Following Butler, mourning can be an agent of transformation, one that resists anthropocentrism and honors the ecocultural interdependencies between people and place. Returning to where we began, with boronia, the very breath of spring, the continued conservation of this species in its habitat south of Perth should also entail the conservation of its historical and cultural networks, as the means by which human entanglement with plants is nurtured and sustained. While the return of boronia fragrance to the city trains and the call of the streetsellers – "Fresh, sweet-scented boronia! Sixpence a bunch, bor-o-nee-a!"(Parker 1962, 4) – might not be possible now, this should not limit us in imagining new human and plant networks that promote interspecies engagements while also ethically managing plant conservation realities. The human-plant future begins with the simple acknowledgement of these networks – the acknowledgement that plants, particularly of the South-West of Western Australia, are a precious and fragile living heritage (a theme explored in chapter 3). And where there is no possibility of biological restoration, where the effects of habitat loss and disease have run their course, memorialization of lost plant species – in literature, theatre, photography, painting and at botanical gardens, environmental centers, government headquarters, homes – can assist communities in mourning their loss.

CHAPTER 2

Toward an Ethics of Reciprocity: Ethnobotanical Knowledge and Medicinal Plants as Cancer Therapies

INTRODUCTION

Turning from the topic of morning and the South-West Australian region discussed previously, this chapter further develops the concept of the posthuman plant through the framework of a reciprocity environmental ethics. Ethnobotanical medicines used in the treatment of cancer will serve as a case study throughout this chapter. The moral virtue of reciprocity, defined as the returning of good when good is received or anticipated, is central to the posthumanist rethinking of human relationships to the plant world. As herbal medicines are used progressively more around the globe and as plant diversity decreases as a result of habitat loss and climate change, an ethics of reciprocity should be a concern for environmental thinkers and conservationists. Aldo Leopold's land ethic and J. Baird Callicott's distinction between deontological and prudential environmental ethics provide theoretical contexts for the development of a reciprocity ethics vis-à-vis ethnobotanical species. While this chapter does not necessarily specify modes or forms of reciprocity, it does outline some of the more prominent ethnobotanical species used in the treatment of cancer, including those from Native American, African, Chinese, and Indian traditions. In the form of a dialogue between the fields of ethnobotany, herbal medicine, and environmental philosophy, this chapter presents a position from which further articulations of reciprocity can be developed, particularly those involving the rights of indigenous cultures and plants.

For a posthumanist relationship to the natural world – one that treats the environment as an active and creative agent to be worked *with* rather than a material, substance, or sight to be worked *over* – the moral virtue of reciprocity is essential. The need for posthumanist perspectives on the natural world is pressingly evident in humanity's interactions with plant life and, more specifically, with medicinal herbs – both cultivated on farms and crafted from the wild. Increased demands have been placed on wild populations of plants to supply herbal medicines that cannot be derived from cultivated species. The number of Americans using herbal medicines between 1990 and 1997 increased

380% (Eisenberg et al. 1998). In 2002, herbal therapy was the leading CAM (complementary and alternative medicine) modality, consumed by 38 million US adults. In 1997, 12.1% of the US population used herbal medicine, whereas, by 2002, this figure increased to 18.6% (Tindle et al. 2005). Moreover, sales of herbal medicines skyrocketed from $200 million in 1988 to $3.5 billion in 1997 and $4.4 billion in 2005. The naturopathic doctor Michael Murray has fittingly pointed to a "herbal renaissance" resulting from advances in pharmacological techniques, increased scientific knowledge of medicinal compounds, and enhanced public acceptance of natural, or complementary, therapies (Murray 2013). Ethnobotanical medicines have been pivotal to the renaissance identified by Murray. Species, such as jimson weed and devil's club discussed later in this chapter, are modern medicines (or plants perceived to have certain medicinal and toxic properties) with documented therapeutic application in traditional cultures. Hence, ethnobotanical knowledge can be characterized through the dialogue or exchange between traditional or indigenous knowledge of plants and contemporary, science-based understandings of herbs and etiology (Waldstein 2014).

Using ethnobotanical medicines from various global traditions for the treatment of cancer as examples, the utilitarian and anthropocentric ethics surrounding therapeutic flora will be analyzed. While several key ethnobotanical species for cancer treatment will be here, it is recognized that an ethics of reciprocity relates to all therapeutic uses of medicinal plants. However, ethnobotanical species for cancer treatment offer salient examples of the need for a reciprocity ethics; conventional medical practices prioritize the alleviation of human suffering, but marginalize the importance of giving back to plants, of returning the favor, in the spirit of reciprocity. The conservation of medicinal plants in the wild ensures an ongoing reservoir of therapeutic plant compounds in the future. But while we consume species of ethnobotanical interest and contribute to (or subtract from) the viability of their habitats, what do we return to the plants from which the medicines have been derived? In contrast to a utilitarian ethics of medicinal plants, the value of reciprocity foregrounds appropriate and sustained exchanges between people and flora that are not based on use-value or virtue-theoretics alone. Leslie Francis defines reciprocity as "the idea of actions-in-return that are not founded in voluntary agreements or contracts" and "doing one's part to produce a common good when – especially because –*others are doing theirs* [italics added]" (Francis 2009). A posthumanist approach to reciprocity recognizes that the term "others" intrinsically includes non-human or "more-than-human" species of flora, fauna, and fungi, not merely humans interacting with other humans (see chapter 1 for more background on posthumanist theory). Moreover, a "common good" is also an ecological good, for the benefit of people, plants, and other beings, involving, among other things, the obligation to respire together.

The taking *from* the plant world should involve a cycle of giving *back* to medicinal species that is not narrowly based in the human attainment of personal health or community well-being. Central to a reciprocity ethics is the understanding, to quote ethicist Francis again, that non-human "others are doing theirs," – that medicinal plants have been offering physically, mentally, and spiritually therapeutic means to humanity (and have been doing so for thousands of years, as indicated by ethnobotanical evidence presented later); and that humanity should do its part, continue to do its part, or invent new ways of reciprocating with plants that ensure the well-being of both – for the intrinsic right-to-exist of both. In *Reciprocity*, originally published in 1986, the American philosopher Lawrence Becker characterizes the virtue of reciprocity as follows: "We ought to be disposed, as a matter of moral obligation, to return good in proportion to the good we receive, and to make reparation for the harm we have done" (Becker 2014, 3). He goes on to classify reciprocity as a "deontic virtue," or a virtue of obligation based on three premises: "We owe a return for all of the good we *receive*, not merely the good we accept;" "the obligations of reciprocity come from the justifiability of being disposed to *make* reciprocation obligatory; [italics in original]" and "the sense of obligation here ought to appear to us, at least in many cases, only in retrospect."

While I accept Becker's first and second premises (returning for good received and being disposed toward reciprocity), I refute a deontic or virtue-theoretic approach over a prudential or act-morality approach. The sense of obligation should and must appear to us not only in retrospect but as a matter of forward-thinking. Reciprocity should and must become part of the moral fabric of our dealings with plants, as part of the foundation of ethnopharmaceutical ventures and conservation initiatives. The urgency of plant conservation demands an act-morality approach, a proactive position of considering the welfare of plants long before harm has been done to them (as subjects-of-a-life) and to their habitats (as communities of abiotic and biotic things). This chapter will enunciate this position in terms of the difference between "should" and "must," using a variety of examples from different locations around the world.

TRADITIONAL ETHNOBOTANICAL TREATMENTS FOR CANCER: GLOBAL CONTEXTS FOR RECIPROCITY

As the world population undergoes the transition to the low mortality and low fertility patterns of industrialized nations, the global incidence of degenerative disease continues to increase. Among these, cancer is especially virulent. In 2012, there were approximately 14.1 million cases of cancer globally (up 225% since 1997), with 7.4 million men and 6.7 women diagnosed. These statistics are expected to almost double by 2035, to 24 million confirmed cases of cancer (Ferlay et al. 2013). In reaction to the increasing global frequency of this disease, a considerable amount of research has been invested in the identification of anti-cancer agents in traditional botanicals. Health research organizations, such as the National Cancer Institute (NCI), have examined the anti-cancer activity of traditional plant medicines as part of a global imperative to address the pandemic, which threatens people in developing and developed nations alike (Balick and Cox 1996). The ethno-medical systems of Native America, Africa, India, and China provide epidemiological leads to plants actually or possibly beneficial in treating cancer, either alone or in combination with other therapies. The value of reciprocity vis-à-vis these species and others should and must be developed as an integral aspect of future complementary health systems that involve plant medicines. By including "should" (suggestive) and "must" (imperative), I emphasize that our moral obligations to plants are both *deontological* (we *must* for the survival of the community) and *prudential* (we *should* for the benefit of ourselves).

Two themes are evident in the literature of ethnobotanical cancer remedies (Martin 2007). The first relates to the identification of plants that have been used traditionally to heal cancer and the relatively direct transmission of this knowledge to the dominant, or allopathic, medical paradigm. However, there usually is no strong correspondence between traditional uses of botanical medicines and their applications in allopathic regimes. Factors such as exposure and predisposition to disease, traditional causes and concepts of disease, introduction of allopathic disease concepts, and life expectancies underlie this lack of correspondence. In general, indigenous terms for "cancer" comprise a variety of conditions including swelling, pain, and malignancies, as the Navajo example will show later in this chapter. The second involves the annexation of botanical remedies, not traditionally known for the treatment of cancer, after the often-inadvertent recognition of potential anti-cancer compounds within the plants. In describing herbs used in cancer therapies, this chapter considers the interrelationships between these themes. In addition to outlining some of the major ethnobotanical remedies for cancer alongside a call for an ethics of reciprocity, this chapter also touches on causal (or etiological) reasoning, particularly the differences (and similarities) between traditional and Anglo-American rationalizations of cancer. The causal agents commonly attributed to cancer in the allopathic paradigm – preservatives, x-rays, radiation, smoking, and sedentary lifestyles – are more-or-less absent from traditional cultures (at least at the time of colonial contact), leading to the issue of differential etiologies and cultural perceptions of disease. For instance, certain food preservatives now banned or no longer widely consumed in some countries have been demonstrated (albeit tentatively) to underlie cancer. This causal relationship is evident in the example of the correlation between stomach cancer and the use of smoke and salt as preservatives (Fontham and Joseph 2005).

Hence, the broader ethnobotanical and etiological questions briefly addressed here include, what is "cancer" in the traditional knowledge systems of some indigenous cultures; and what causes it? How do traditional perspectives of cancer influence the prescription or administration of ethnobotanical remedies? How can the ethics of reciprocity advocated here enhance the efficacy of ethnobotanical species, while at the same time acknowledging the intrinsic right-to-exist of the plant and the imperative to "return good in proportion to the good we receive" from the botanical world? The next section explores the former question in terms of environmental ethics.

PLANTS AS SUBJECTS-OF-A-LIFE: INDIVIDUALISTIC, HOLISTIC, AND GAIAN APPROACHES

In order to develop a reciprocity ethics, I first will discuss the criticism that certain modes of environmental ethics privilege single organisms or species of organism—including human beings and ethnopharmaceutical plants. Indeed, the reciprocity ethics I am calling for tends to focus on giving back to individual medicinal species, such as ginseng and devil's club. Is an individually-based ecological ethic *really* environmental, sustainable, or ethical at all? This provocation brings into focus the ideological rift between holistic and individualistic models that is evident in the literature. On one side, individualists aver that the principle of "subjects-of-a-life" (Regan 1993) ethically distinguishes living beings (insects, bacteria, plants) from non-living things (rocks, soil, detritus) that are not rights-possessors. The ethical line drawn between species can create a hierarchy of environmental values in which ginseng and devil's club are privileged over the abiotic environment of which they are part. The positions of Peter Singer and Tom Regan, for example, express a concern for the welfare of individual ecosystem members, with particular attention paid to large, charismatic, sentient, and ostensibly intelligent mammals (and I would argue charismatic flora such as old-growth trees). In contrast, the holistic environmental paradigm, based in Aldo Leopold's land ethic, asserts an inherent relationship between the rights of the "land" (collectively including biological processes involving soils, plants, and animals) and the welfare of individual beings. In Leopold's model, the individual freedoms of subjects-of-a-life can sub-serve the rights of the ecosystem, especially when community interests are at stake. In other words, Leopold's model is a community-based environmental ethics in which land is conceived of as a society of beings and non-living things.

However, I suggest that the ethical structure underlying the holism of Leopold and also of the philosopher J. Baird Callicott – the correlative relationship between individual rights and the "integrity, stability, and beauty of the biotic community" (Leopold 1993) – is not entirely at odds with individualist models. Callicott builds on Leopold's holistic precepts, arguing that the land ethic is "self-consistently both [...] deontological [and] prudential" (Callicott 1993); in other words, the reciprocal relationship between the natural world and the human community (indeed sustained by a land ethic) can ensure the welfare of both. In this context, the term "deontological" refers to human obligations, rules, and duties to the natural world; and is closely associated with virtue ethics and "doing one's part." "Prudential" refers to the good of the individual or the subject acting on behalf of itself. However, Leopold and Callicott equally fall short of articulating reciprocity as an environmental value, although the ethical tenet is implicit in their models, particularly Callicott's. As a counter-example to notions of reciprocity in holistic environmental ethics, the Gaia theory contends that the intrinsically self-regulating processes of the Earth – its organismic qualities – could give rise to new ecosystems, not necessarily including human societies, in response to the disturbance of the biosphere by climate change and other potentially catastrophic ecological issues (Schneider 2004). By theorizing the homeostatic self-regulation of the Earth, Gaian theory seems to abnegate the place of environmental ethics of both kinds (holistic and individualistic) by ruling out the possibility of reciprocity (or mutual benefit) between the land and individual beings or species of beings, as discussed later in this chapter.

In contrast to Gaian theory, Leopold's land ethic has the potential to encompass reciprocity as an act-morality principle. Rather than the magnum opus of "environmental fascism" (Regan 1983, 361–362), the land ethic appeals to individual self-realization through reciprocal, dialogic engagement with an ecologically functional and intact community of beings and natural things.

Therefore, healing plants are subjects-of-a-life as well as members of ecological (and social, cultural, economic, medicinal) communities, comprising human and non-human beings. The more that environmental ethical structures can avoid hierarchies and taxonomies, the more true ecological justice can be realized. And this is not merely blue-sky ideation. Our moral obligations to plants are both deontological (*we must* for the benefit of the community) and prudential (*we should* for the benefit of ourselves). It is only in reference to a land-as-community model (in Callicott's sense, building on Leopold's) that an ethics of reciprocity can be viable because reciprocal exchanges *must* be inclusive, thus involving subjects-of-a-life not traditionally encompassed within eco-ethical frameworks. These subjects include plants and, more precisely for this discussion, medicinal herbs, such as poke, jimson and devil's club of indigenous North American traditions, discussed in the next section. In summary, the position I am arguing for recognizes plants as subjects-of-a-life (extending Singer and Regan's work on animals) as well as plants as members of land-as-community (reflecting Leopold) while rejecting an ecological ethics based in Gaian theory. The subject-of-a-life is a priori a subject of an ecological community.

Poke, Jimson, and Devil's Club: Traditional Native North American Cancer Botanicals

Indigenous North American nosology (disease classification) indicates the prevalence of rheumatism and arthritis, dysentery and other digestive disorders, intestinal worms, and eye disorders in Native American populations at the time of European contact. Modern diseases, such as cancer, heart disease, and arteriosclerosis, however, were uncommon (Vogel 1970, 161). Nonetheless, traditional ethnobotanical treatments for cancers and tumors existed and have been documented, as indicated by the following examples of poke, jimson weed, and devil's club.

American pokeweed (*Phytolacca americana* L.) is a pungent perennial herb of the family Phytolaccaceae, and is native to North America in dry fields, hillsides, and along roads. Its many folk names include Virginia poke, American nightshade, cancer jalap, coakum, garget, pigeon berry, pokebush, pokeberry, pokeroot, poke sallet, inkberry, ombu, redweed, and scoke. Indigenous North Americans used poke roots and berries medicinally. The large, mature root can be easily broken and sliced. A thin, brownish bark covering the fleshy and fibrous root tissue can be readily peeled. Dark purple, globular berries ripen in late summer and autumn in the northern hemisphere. Although poke has a long history of efficacy in numerous indigenous North American medical systems, its toxicity in large amounts should caution potential modern users or experimenters (Hutchens 1991, 223–224). For example, in traditional Cherokee medicine, an infusion of poke berries was taken for arthritis and rheumatism (Hamel and Chiltoskey 1975), whereas for the Delaware people, roots were used for glandular swellings and chronic sores, and to purify the blood (Tantaquidgeon 1972). In nineteenth-century North America, poke poisonings were documented, particularly cases involving the misapplication of the tincture as an anti-rheumatic and from the mistaken ingestion of the toxic berries. However, a tincture of the fresh root harvested in winter and a tincture of the ripe berries have been shown to be potentially helpful in the treatment of different cancers. In particular, anecdotal evidence corroborates the efficacy of poke root, and indicates that Native Americans used the powdered root as a poultice for "cancerous ulcers" (Hutchens 1991, 351). Modern herbalism draws from the traditional usage of pokeweed root, ground fine, and applied as a poultice, most notably for the treatment of breast cancer (Dominion Herbal College 1969, 13). Grated pokeroot can be applied to the breasts to treat inflammation and rashes. Moreover, research has shown that pokeweed antiviral protein (PAP) has antitumor properties in laboratory studies. Other studies have demonstrated that PAP can be used to treat advanced osteosarcomas and soft tissue sarcomas when combined with immunotherapy drugs (American Cancer Society 2008). A more recent study (2014) indicates that alcoholic extracts of pokeweed change the expression of genes associated with colon cancer, potentially enhancing treatment (Maness et al. 2014).

Jimsonweed (*Datura stramonium*), or datura, is a species within a small genus of twelve species of shrubs or annual or perennial herbs, belonging to the nightshade family, Solanaceae. Jimsonweed is native to North America, an annual, over four feet tall when mature, with ovate, unevenly-toothed, glabrous, and pungent leaves; white or purplish funnel-form flowers; and a hard, barbed, multi-seeded capsule. Its numerous and often suggestive common names include thorn apple, green dragon, hell's bells, devil's trumpet, devil's weed, tolguacha, Jamestown weed, stinkweed, locoweed, prickly burr, devil's cucumber, and sacred datura. The name "jimsonweed" is a corruption of "Jamestown weed," supposedly derived from an incident during Bacon's Rebellion, an armed uprising of Virginia settlers in 1676. While jimsonweed is used as a medicine in Asia, where it was introduced, other datura species are native to the Old World. Although almost every part of the plant possesses medicinal properties, the most regularly used are the leaves and seeds. In large amounts, jimson weed is an energetic narcotic poison and is seldom prescribed by contemporary herbalists, or is used with extreme care in low doses (Dominion Herbal College 1969, 166).

Despite its dangers, jimsonweed has a wide range of traditional uses, in North America and Asia in particular. The smoke from the burning leaf is inhaled for bronchitis and asthma. The juice of the berries can be applied for dandruff and scalp disorders. Seeds and leaves are known to possess sedative properties and, thus, have been used to treat hysteria, psychosis, and insomnia. The compounds scopolamine, hyoscyamine, and atropine have psychotropic effects, underlying the use of jimsonweed for hallucinatory purposes in some traditions (Schultes, Hoffman, and Rätsch 2001). The Lumbee Indians of the Mississippi River report jimson weed as an external application for cancer, presumably as a poultice of the ground root or berries (Croom 1992). Moreover, anecdotal accounts confirm the application of seeds and extracts (tinctures of the root, for example) for a variety of acute afflictions, including "ulcerous affections and cancer" (Vogel 1970, 327). Alkaloids, tannins, carbohydrates, and proteins in jimson are being screened by medical researchers as beneficial compounds in cancer treatment. The protein Datura Stramonium Agglutinin (DSA) has been isolated from jimson weed as a possible treatment for malignant gliomas of the brain (Cancer Research UK 2013). Japanese scientists found that DSA entirely impeded the growth of cancer cells in a laboratory. When DSA was applied, the cells differentiated and shed their malignant properties. In this study, DSA was shown to induce the differentiation of glioma (tumor) cells, potentially offering a therapy for treating certain forms of cancer without the side effects of chemotherapy (Sasaki et al. 2002).

The last ethnobotanical example from the indigenous North American tradition briefly presented in this section is devil's club (*Oplopanax horridus* (Sm.) Miq.), a member of the Araliaceae family. Related to ginseng, devil's club possesses general tonic, or adaptogenic, properties. Its other common name is devil's walking stick. A widespread species in north-west British Columbia, devil's club is a deciduous shrub with a sprawling habit, three to fifteen feet high, and growing in damp evergreen and mixed forests. Native American communities have long known devil's club to control diabetes (Yance 2013). The Gitksan (or Gitxsan) of British Columbia gather the leafless prickly stems and usually scrape off the inner cambium of the stems after senescence or when the plant is dormant. The inner bark of devil's club is applied dried or fresh for cancer, especially gynecologic, and for stomach ulcers (Johnson Gottesfeld and Anderson 1988). In addition to its use in Gitksan culture, devil's club has been gathered as a cancer treatment in the Tlingit and Tsimshian cultures of north-west British Columbia

(Johnson Gottesfeld and Anderson 1988, 20). Recent medical research indicates that devil's club inhibits the growth of several forms of cancer. A study found an extract from the plant (OhE) effective for alleviating human colorectal cancer. OhE has also been shown viable as an ovarian and breast cancer therapy (Li et al. 2010).

LAND ETHIC AS FRAMEWORK FOR RECIPROCITY: DEONTOLOGICAL AND PRUDENTIAL REGARD FOR PLANTS

The reciprocity framework that I am proposing, vis-a-vis medicinal plants, such as poke, jimson weed, and devil's club, should be developed in relation to a holistic environmental ethics that is both *deontological* (the good of the commons) and *prudential* (the good of the self), to again borrow J. Baird Callicott's terms. The land ethic, formulated by the ecologically-versed and poetically-aware Aldo Leopold, certainly extends beyond him, past the Dust Bowl days of the American 1930s, and connects to the extant traditions of indigenous North American peoples, such as the Gitksan and Algonquian, who have developed complex traditions of giving back to the natural world in exchange for *taking from* (Callicott 1993, 127). Indeed, indigenous models of reciprocity would have influenced Leopold's vision of cooperation and interdependence between human individuals and the land as a community of beings. The sustainability of traditional cultures across the world and over the millennia (for example, the fifty-thousand-year cultural traditions of the Nyoongar of South-Western Australia; see chapters 4 and 9) is compelling indication of reciprocity between human communities and their local ecosystems. Exchange is prudential and intrinsic to the long-term welfare of cultures and beings (Rose 1996). However, it should be noted that longevity does not always correspond to ecological reciprocity. For example, environmental and cultural influences might maintain low human populations in which anthropogenic impacts are negligible, despite the intensity of activities by individuals or groups. Leopold reflects an appreciation of these traditions and exigencies in his writing (although his Darwinian side shines through), characterizing an ecological ethic as "a limitation on freedom of action in the *struggle* for existence [italics added]" (Leopold 1993, 95). In particular, he extends the reach of ethical concern to the land by thinking across the terms "politics and economies" (and thus about human social communities) and "symbioses" (and thus about ecocultural reciprocity) to illuminate the premise of interspecies relations linking both: "Politics and economics are advanced symbioses" (Leopold 1993, 96). In Leopold's view, extending the *logos* of human ethics, the land ethic, to some degree, attempts to constrain the behavior of individuals in order to limit "free-for-all competition" (Leopold 1993, 96). In other words, humans would not wantonly destroy human life; why should we cut down tracts of forest, drain wetlands areas, and exploit medicinal plants through overharvesting, with no concern for the plants themselves or for the future of humanity's medicine?

However, the subjugation of individual rights to ecological oligarchy is not the core of Leopold's argument, though it might appear so on superficial reading. Instead, Leopold contends that, in order to benefit individuals, a community ethic should advocate land protection, which will return good to the individual in shared ways (for example, through the provision of clean air, food, and water, as well as spiritual rejuvenation in unspoiled places). In the land-as-community model, nature has both instrumental and intrinsic value; the former cannot be divorced from the latter, and both confer mutual advantage. Leopold alludes to the tacit effects of the land ethic on the individual: "An ethic may be regarded as a mode of guidance for meeting ecological situations so new or intricate, or involving such deferred reactions, that the path of social expediency is not discernible to the average individual" (Leopold 1993, 96). The principle of "deferred reaction" is intrinsic to Leopold's thesis. Yet I suggest that this protracted effect, whereby the advantages to human communities are not immediately apparent, underlies the dichotomy between individualistic and holistic environmental ethics. In other words, the current of ethical concern for the community need not drive the individual to thirst, nor would Leopold have it this way. The land ethic is not a misanthropic position. Instead, it suggests that the community affects what is determined to be advantageous to individuals, and that which is suitable for the long-term welfare of ecological societies, including people and plants. As part of this reflexivity between the community and the individual, an ethics of reciprocity is intrinsic and signifies that all are doing their parts. Medicinal plants provide therapeutic compounds and spiritual sustenance (especially in the case of plants such as sacred datura) and human beings return the materials and modes of regard necessary for the long-term wellbeing of the plants (as subjects-of-a-life) and their habitats.

PSOROSPERMUM, CASAVA, AND PERIWINKLE: TREATING CANCER IN TRADITIONAL AFRICAN ETHNOBOTANY

In the previous section, I suggested that reciprocity is complementary (and arguably essential) to Leopold's land ethic; and that ecological ethicists ought to consider the moral implications of overusing, exploiting, commodifying, or driving to extinction plants used for human health and healing. This section goes on to detail other plants of relevance to the treatment of cancer, both historically and in contemporary allopathic contexts. Numerous plants used in African traditional medicine have been investigated for their cytotoxic (toxic to cells in higher doses but possibly therapeutic in lower doses) and antineoplastic (acting to prevent or inhibit neoplasms or tumors) properties. The three examples that follow – *Psorospermum febrifugum* Spach, *Maprounea africana* Müll.Arg., and *Catharanthus roseus* (L.) G.Don – will provide some indications of the potential of African plants to supply anticancer treatments in a world increasingly affected by cancers of different types and in a world in which wild plants are ever more threatened. However, I emphasize that the following three plants are not only the materials (chemicals, compounds, agents, substances) used for human health, but are first and foremost subjects-of-a-life in themselves, beyond their utilitarian applications and potentialities. Hence, the fields of medical herbalism and ethnobotany need to think beyond the use-value of plants and toward the subjectivities of the species involved, that is, toward plants as subjects-of-a-life with specific rights to exist in their original habitats.

Traditional African healers employ *Psorospermum febrifugum*, a bush found over wide areas of central and eastern Africa, including Senegal, South Sudan, Ethiopia, Mozambique, Zimbabwe, and Angola, to reduce fevers, as the species name indicates (*febri-* for fever). *Psorospermum* is in the Clusiaceae (or Guttiferae) family, consisting of thirty-seven genera and over one-thousand six-hundred species distributed mainly in tropical areas. While investigating the antipyretic properties of the herb, researchers also identified anticancer compounds. Through an in vitro bioassay, scientists isolated from the roots of *Psorospermum* an antitumor and antiviral form of the compound xanthone called "psorospermin," which is active in controlling mammary and colon tumors. Another compound called an "anthrone" has been demonstrated in mice to possess *in vivo* activity against P-388 leukemia (Hostettman and Marston 1986). Moreover, *Psorospermum* extracts were found to be effective against A2780cis ovary cells, a malignant cell type that is resistant to the anticancer drug cisplatin. Omodin is an antitumor compound identified in the species as valuable for treating lung, prostate, ovarian, colon, and hepatic cancers (Tamokou et al. 2013).

Magic nut, redskin bush, or mburabu (*Maprounea africana*) has been used in traditional eastern African cultures as a purgative and to cure syphilis. Although sometimes referred to as tree cassava, *M. africana* should not be confused with *Manihot glaziovii*, an Amazonian species introduced to Africa and also known by this common name. Magic nut is a deciduous shrub or small tree with hanging branches and reddish-brown twigs and reddish-yellow flowers in heads. The species occurs in Benin, Tanzania, Angola, Namibia, Botswana, Zimbabwe, and Mozambique. Its bark is taken as a purgative in low doses, with larger amounts being highly toxic and potentially fatal. A decoction of the root is consumed traditionally to alleviate syphilis, venereal diseases, leprosy, and dysentery. Preparations of the roots, bark, and leaves are employed in Gabon as a diuretic. In Congo, stems and leaves are chewed for constipation, intestinal worms, and irregular menstruation (Schmelzer 2008, 376). Alcoholic extracts of the dried roots exhibit activity against p-388 leukemia in mice. Further research has led to the identification of a number of pentacyclic triterpenes, one of which is highly active in the p-388 in vivo test (Hostettman and Marston 1986, 121).

Indigenous people in Madagascar have been using rosy periwinkle (*Catharanthus roseus*) as a treatment for insect stings, eye infections, toothaches, malaria, diabetes, and cancer. The species is also known as vinca, Cape periwinkle, and old maid; and is endemic to Madagascar although it is found elsewhere, including Jamaica. Periwinkle is an herbaceous plant with oblong leaves and white to dark pink flowers with red centers. In 1958, Gordon Svoboda screened a periwinkle extract, later identifying over seventy alkaloids in the plant. In fact, specimens of *Catharanthus roseus* were collected in Jamaica for use in diabetes trials, but the results were inconclusive. Attempts to verify the folkloric use of periwinkle as a diabetes treatment later led seredipitously to the identification of two alkaloids applied in the clinal treatment of cancer (Noble 1990). These alkaloids, vincristine and vinblastine, are now common around the world in treating pediatric leukemia and Hodgkin's disease (Balick and Cox 1996, 33). The alkaloids inhibit the division of cells in lymphomas, leukemias, and tumors. While these chemotherapy drugs are now well established as important treatments for various types of cancer, none of the benefits of over fifty years of commercialization have been shared with indigenous people of Madagascar while periwinkle itself becomes gradually more endangered in its native habitat (Kiene 2011, 16). Although its status in the wild is compromised, because of the impacts of agricultural practices, the species is extensively cultivated outside of Madagascar. This brief example demonstrates a highly utilitarian, capitalistic, and exploitative approach to ethnobotanical species in which reciprocity is not figured into modes of exchange between the plant, local indigenous people, and the pharmaceutical industry.

ETHICS AS COMMUNITY INSTINCT: LOVE AND RESPECT FOR
MEDICINAL PLANTS

Concepts of reciprocity between land and organism, between community and
member, between common energy and individual resources, are nascent within
Leopold's land ethic. Regarding his marginalization of the role of the individual,
J. Baird Callicott in "The Conceptual Foundations of the Land Ethic" comments
that most philosophers have regarded the land ethic "with horror because of its
emphasis on the good of the community and its de-emphasis on the welfare of
individual members of the community" (Callicott 1993, 125). I suggest that part
of mainstream philosophy's "horror" is due to the non-dualistic nature of
reciprocal engagement with the land and other living beings, in which the I-thou
distinction dissolves, in which the highly individuated subject becomes an
unstable ontological state. In a paradigm of reciprocity, self-interests cannot be
achieved through self-dialogue, through a form of environmentally dangerous
solipsism. For Leopold, concepts of community and ecosystem embed the
individual. For Callicott, the ecosystem provides the context of individual
assertions, but, more specifically, genetic predeterminations underlie our feelings,
instincts, and capacities – for "love, sympathy, respect" (Callicott 1993, 131) –
nurturing the developmental processes and maintaining the community as the site
of all ecosocial interactions. Hence, for Callicott, love and respect for the natural
world are as important to an ethics of reciprocity as they are to civil society.

Extending Leopold's notion of deferred social expediency to the evolutionary
origin of community ethics, Callicott concludes, as stated previously in this
chapter, that the land ethic is both deontological and prudential. As "a kind of
community instinct in-the-making" (Leopold 1993, 203), ethics is a deontological
appeal to community demands through values of duty, self-sacrifice, love, and
respect for non-human beings, including ethnobotanical species, such as
periwinkle and tree cassava, and their broader habitats. Genetically influenced in
Callicott's view, the frameworks of environmental ethics ensure community good
by constraining purely personal gain (read: the small percentage of mining
magnate billionaires in Australia) where the good of the whole becomes,
reciprocally, the good of the individual (read: to live a life of quality, unrelated to
the greedy accumulation of capital). But the land ethic is ultimately also
prudential because ethical treatment of land returns good to individuals,
particularly in the form of ecological good (clean air, food, water). These
dynamics between prudential concern and deontological regard for the
environment are best summarized in Callicott's statement that "'there is no way
for land to survive the impact of mechanized man [*sic*]', nor, therefore, for

mechanized man [*sic*] to survive his [*sic*] own impact upon the land" (Callicott 1993, 132). In the context of mechanized allopathic health care that looks to integrate (and, in many instances, exploit) ethnobotanical traditions, such as those previously described, survival in all senses depends on land, despite the carrying out of laboratory investigations in sterile spaces and the illusion of detachment from ecological exigencies during experimental processes. Love, respect, and reciprocity need to underlie all interactions with ethnobotanical species and with the indigenous people who have safeguarded them for thousands of years. This returns good to the plants in their environments commensurate to the good we receive from them in the form of medicinal compounds, community well-being and, perhaps most importantly, the realization that we are not alone on this planet, that we are always part of ecocultural communities.

ASHVAGANDA, GINSENG, REISHI, AND LICORICE: AYRUVEDIC AND CHINESE MEDICINAL PLANTS

Ayurvedic and ancient Chinese medical systems have well-established traditions of using plants as therapeutic agents for a variety of ailments. Ayurvedic treatments for cancer exhibit a constitutional basis, in which a unique remedy is determined according to the alignment of the patient or condition to one dosha or a combination of doshas: vata (air), kapha (earth), and pitta (fire). Robert Svoboda's book *Ayurveda: Life, Health and Longevity* describes the case of Dr Agate, a professor in an Ayurvedic college, who was diagnosed with advanced stage acute myeloblastic leukemia (AML) and given no longer than one year to live (Svoboda 1992, 299–303). Ignoring blood reports, bone-marrow tests, and the overall allopathic diagnosis, he focused on Avurvedic therapies, in which the intense pain in his bones and joints was attributed to the vata dosha and wind invasion. Agate carefully followed a regime of herbs prescribed by a consulting Ayurvedic physician. The treatment took the form of a powder containing ashvaganda (*Withania somnifera* L. Dunal), sariva (*Indian sarsparilla*, *Hemidesmus indica* R. Br.), and amalaki (Indian gooseberry, *Emblica officianalis* Gaertn.). The blast cells in Agate's blood began to return to normal after about six months and eventually he could resume his teaching duties.

Of the three herbs used by Agate, ashvaganda exhibits scientifically documented anticancer properties (Svoboda 1992, 298–301). Regarded as the primary Ayurvedic strengthening tonic, or an adaptogen like ginseng and Devil's club, ashvaganda is a perennial herb that grows to a height of five feet with a width of about three feet. The stem is green and erect; the leaves are ovate, green, and alternate; the flowers are small, greenish white with a white stigma; and the red berries are encased in a papery sheath. With sedative and narcotic properties, ashvaganda is a widely used herb in Ayurvedic medicine. It has documented antitumor activity, largely attributed to the major chemical component, withanolides (steroidal lactones), including somiferin and withaferin A (Patel 1986). Ashvaganda extracts have been demonstrated to increase platelet, and red and white blood cell, counts during cancer chemotherapy treatment with cyclophosphamide. Moreover, animal studies in India conclude that ashvaganda sensitizes cancer cells to radiation therapy, making treatments approximately fifty percent more efficacious. Studies have shown that ashvaganda facilitates the regression of cancerous tumors (Balch 2002, 26).

As in Ayurveda, the traditional Chinese botanical treatment for cancer focuses on tonifying the whole body, not just the afflicted organ or system. Only in the last few decades has the efficacy of various Chinese herbs been subjected to allopathic methods of scrutiny, such as screening tests. In particular, Chinese herbal medicine exemplifies the role of traditional botanical tonics in the contemporary treatment of cancer. Ginseng (*Panax ginseng* Meyer) has been central to Chinese medicine for over two thousand years as an aphrodisiac, painkiller, and general stimulant, although American ginseng (*Panax quinquefolius* L.) is even more highly valued. *P. ginseng* is a low-growing perennial plant, native to regions of China, North Korea, and Siberia. In the autumn, ginseng root is dug, washed, steamed, and dried for use. A study in South Korea observed that individuals who habitually consume ginseng have a sixty percent lower incidence of death from cancer, especially of the lungs and stomach. Additionally, research in China found that when ginseng therapy was combined with traditional radiation and chemotherapy for small cell lung cancer, the patient's life span increased by three to seventeen years. Moreover, polyacetylinic alcohol in ginseng impedes tumor cell reproduction and augments the effectiveness of the drug mitomycin in stomach cancer therapy. A study of almost two thousand individuals concluded that regular ginseng use reduces the likelihood of developing many forms of cancer (Balch 2002, 74).

Reishi (*Ganoderma lucidum* (Fr.) P. Karst) is regarded as an "elixir of life" in Chinese medicine and is increasingly known around the world as a potent tonic for energy, disease resistance, and longevity. Although over ninety-nine percent of all wild reishi mushrooms are found growing on old plum trees, fewer than ten mushrooms will be found on one-hundred thousand trees, lending the reishi its common name "phantom mushroom." Some of its other vernacular names are lingzhi and king of herbs. Reishi has numerous therapeutic applications, for example, anticancer, immunoregulatory, antioxidant, liver-protecting, hypoglycemic, antibacterial, antiviral, antifungal, and blood cholesterol lowering effects. Reishi activates the body's production of interleukin-2, which protects against several kinds of cancer, and contains ganoderic acids, which act against liver cancer. Reishi counteracts the suppression of red and white blood cells that can result from cyclophosmamide treatment by stimulating the production of bone marrow protein (Balch 2002, 116). Another study of reishi concluded that the fungus suppresses the adhesion and migration of invasive prostate and breast cancer cells, suggesting its usefulness in reducing tumor development of these kinds. The anticancer properties exhibited by *Ganoderma* underscore its potential as a dietary supplement in conjunction with other alternative therapies for cancer treatment (Sliva 2003).

The final plant discussed in this section, licorice (*Glycyrrhiza glabra* L.), has been used for over three thousand years in traditional Chinese medicine as a tonic to rejuvenate the heart and spleen, and as a treatment for ulcers, cold symptoms, and skin disorders. Also known as sweet root, licorice contains sugar-like compounds and has been used for a range of ailments, particularly as a demulcent and expectorant. Native to parts of Europe and Asia, licorice is a woody-stemmed perennial that attains a height of six feet, bearing clusters of creamy white flowers. In the autumn, roots of three- to four-year-old plants are dug up. Licorice protects the body against a range of carcinogenic compounds, including chemotherapy toxins. *Glycyrrhiza* prevents the formation of skin tumors caused by noxious chemicals. Additionally, licorice hinders the cancer-causing effects of pollutants, such as benzopyrenes, and a chemical called aflatoxin that results from improperly stored food grains. Licorice also defends the body against some arsenic compounds, urethane, caffeine, and nicotine (Balch 2002, 91). Recent research indicates that licorice slows the growth of skin cancer cells by blocking the proteins required for the development of melanomas. Another chemical in licorice, isoangustone A, has properties similar to glycrrhizin but without the associated side effects of arrhythmia, high blood pressure, and muscle weakness (The Huffington Post UK 2013).

GAIA THEORY AND AN ETHICS OF RECIPROCITY: CAN THERE BE AGREEMENT?

The principles underlying Leopold's land ethic and the community ethics of Callicott imply the value of reciprocity between community and individual. Yet reciprocity as a principle is not explicitly developed in their arguments. Making reciprocity explicit and providing a theoretical framework for doing so have been the focal points of this chapter. Despite the criticism that the land ethic marginalizes the individual for the welfare of the whole, the holistic environmental ethics of Leopold and Callicott, in fact, implicitly attend to the well-being of both the individual and the community – the organism and ecosystem. An individualism enhances the community (i.e. the land, habitat, biological system) while a collective focus strengthens the constituents (i.e. the subjects-of-a-life). This is the context for reciprocity – in which acts of contributing to and giving back (or doing one's part) can take place. In other words, the reciprocity ethics I am calling for is both species- and land-based.

In contrast to a land ethics, the Gaia hypothesis developed by Lovelock and Margulis theorizes the Earth as a homeostatic system that alters its ecological processes—soil and atmospheric composition, and floral and faunal make-up – in response to human interventions. In their view, homeostasis (involving broad scale temporal and geological phases) is a normative feature of the Earth's ontology. The planet regulates factors of climate and temperature in order to establish suitable conditions for community members – human and non-human, biotic and abiotic. Gaia theory calls attention to life facing decline in the event of a global equilibrium shift, triggered, for instance, by climate change. In this framework, planetary life systems, perpetually in flux and adapting to new conditions, are not contingent on conditions of reciprocal exchange over time. In fact, Lovelock considers ideas of planetary stewardship to be ridiculous and dangerous "hubris." Since we will never know enough about the complexities of microscopic life – in his view, the basis of the Earth's life systems – an ethical approach is a hands off one that minimizes human involvement and, therefore, seems to abnegate ethical responsibility (Lovelock 1988, 206). A Gaian perspective counters the position that human life and land-as-community are bound to reciprocal engagements. Humanity ultimately bear the consequences of ecological myopia and acts of reciprocity cannot turn the tide.

In another sense, the extreme holism of Gaian theory can lead us full circle to individualistic concerns. If the consequences of global climate change are definitive, then a course of action is to re-emphasize the place of the individual, in order to protect ourselves from radical ecological shift. This emphasis on the individual suggests that the community and its members are interdependent, as Leopold and Callicott would have it. In Gaian theory, ethical frameworks are constructions having no real bearing on the planet's wellbeing. If the planet's feedback mechanisms involve a constant state of flux over the millennia, how can we define or isolate a baseline state of ecological wellbeing from which ethics can proceed? How can we determine land health in the context of Gaian theory in which our limited human vantage point is a stark and isolated one within the expanse of the Earth's geological history? The breakdown of notions of reciprocity in light of Gaian theory is in contrast to the land ethic, where collective concerns of survival, autonomy, and self-realization play out. Gaian theory is at odds with the land ethic, although both are often erroneously subsumed within the heading "holistic ethics." Moreover, individualistic philosophies (in my view, not sufficiently articulated by ethicists such as Singer and Regan) and the land ethic (of Leopold and Callicott) are less in opposition than they appear, or at least should be regarded as less antagonistic if genuine practices of reciprocity are to be realized.

The holistic ethics of Leopold and Callicott – rather than Gaian theory – offer an amenable position from which to advance notions of reciprocity in relation to the land ethic and medicinal plants. When an ecosystem fails, so does human and botanical wellbeing alike. This exigency touches on the core of contemporary environmental problems: the disruption of the land community (through pollution, deforestation, loss of biodiversity) is intimately linked to the decline of the individual organism (through air quality depletion, disease, loss of hope). In other words, an ethics of reciprocity concerns the whole (*i.e.* earth, communities, systems) in dialogic exchange with the constituents (*i.e.* subjects-of-a-life). As Callicott argues, this position mediates prudential self-interest and deontological concern for community. In recognizing plants as subjects-of-a-life, we come to see them also as members of land-as-community. A reciprocity ethics is one that returns good to plants and their broader biocultural environments, despite Gaian contexts of climate change which can disempower such a position. My exploration of ethnobotanical treatments for cancer affirms the following point: the receiving of ethnobotanical good should be balanced by a giving back of good to the plants themselves, the environments in which they grow naturally, and the indigenous people whose cultural heritage involves medical knowledge of the species. It is not enough to privilege cultivating healing plants as a solution to their disappearance in the wild. As species decline, the ecocultural knowledge systems associated with them become at risk, as the next section goes on to explain through the example of the Navajo experience of cancer.

CANCER, CAMAS, AND CROTON: TRADITIONAL ECOLOGICAL KNOWLEDGE AND RECIPROCITY WITH PLANTS

Reciprocity with the botanical world also entails reciprocity with indigenous peoples who have maintained traditions of healing with plants for centuries and whose ecological knowledge often informs (and is exploited by) medical science. Thus far in this chapter, traditional botanical healing agents for cancer have been examined in reference to some modern research that has either expanded on their traditional uses or disclosed previously unknown anticancer properties. However, the cause, meaning, and identity of cancer in traditional knowledge systems underpins the use of ethnobotanical curatives. An example of Navajo conceptualizations of cancer will provide some insight into cross-cultural etiology and nosology. For example, two patients in a study claim to have been definitively cured of cancer, one by traditional herbs and one by peyote (*Lophophora williamsii* (Lem. exSalm-Dyck) J.M. Coult). In the study, forty-three percent of traditional treatments used by Navajo cancer patients involved Navajo herbs administered in the Lifeway, Pus-eater, or peyote ceremonies. As indicated by the term "Pus-eater," Navajo thought conceptualizes cancer as a putrescent sore, rather than as a growth or tumor: "Negative, uncontrolled growth is a less culturally salient metaphor for Navajos than for ourselves. In Navajo thought, growth is inherently positive, whereas degeneration and decay are characteristically negative processes [...] To conceive of cancer as something that 'keeps on rotting' is more consistent with such a view, while our own conceptualization of 'unchecked growth' is consistent with our fear of nature (and society) out of control" (Csordas 1989). Navajo etiology accordingly describes cancer by using a vocabulary of decay, rather than one of negative growth, as common to the allopathic perspective.

Instead of a basis in syndrome and symptom, the Navajo disease classification system centralizes etiology, or causation. In contrast to the causation cited by Anglo-American participants, Navajo patients ranked lightning as an important etiological category. To the Navajo, lightning is more than a cosmological and an environmental reality. It is a metaphorical fact of life: the category of lightning extends beyond the storm-related kind to include other forms of radiant energy. For example, radiation from a uranium mine and exposure to the flames and fumes of a welder's torch were cited by Navajo patients as likely causes of their illness. The study suggests that a physical or bodily causes (injuries) ranks significantly alongside a spiritual or celestial causes (lightning). The Navajo conceptualization of cancer provides a brief insight cross-cultural etiology, but most importantly, the example underscores the capacity of traditional ethno-medical systems to encompass and adapt to modern disease classifications. Lightning is an archetypal form of radiation, but radiation is also a modern Navajo interpretation of the traditionally broad etiological category of shooting phenomena, including snakes and arrows (Csordas 1989, 465).

For indigenous cultures such as the Navajo, an ethic of reciprocity is closely related to medicine and food. As such, notions of self-interest are not generally reflective of individual behaviors in which deontological concerns balance purely prudential ones. Historian Adam Sowards in his book *United States West Coast: An Environmental History* (2007) opens the chapter "Reciprocity and the Indigenous Landscape" with a scenario of a Native American woman of the Pacific Northwest using a digging tool to search for the blue flowers of camas (*Camassia quamash* (Pursh)) (Sowards 2007, 19). Also known as Indian hyacinth or wild hyacinth, camas species were important food sources for Native American peoples and early settlers. The bulbs were roasted or boiled, tasting like sweet potatoes, or pounded into flour and stored. In Sowards' anecdote, the woman and her companions prepare to dislodge the bulbs of the season's first camas from the earth. In return for the nutritious bulb, the women offer their tobacco and prayers, asking permission to harvest before proceeding to do so. They also returned the flowering stalk of the plant to the earth. Some of the camas were stored away for winter ceremonies. Others were boiled and mixed with a sweetener to prepare a cough medicine. This ritualized harvesting of a plant involves taking as a form of borrowing from and giving back to the earth.

A modern example of reciprocity as an ethic comes from Shaman Pharmaceuticals, a defunct company, later reincarnated as Napo Pharmaceuticals, that has bioprospected traditional knowledge from indigenous healers and herbalists to create FDA-approved ethno-pharmaceuticals. The company commits a percentage of its profits to the indigenous communities from whom they have acquired ethnobotanical intellectual property. Shaman also founded the Healing Forest Conservancy, a nonprofit organization aimed at compensating indigenous peoples by preserving cultural and biological heritage (Hefferon 2012, 21–22). The drug crofelemer is derived from sangre de grado or sangre de drago (*Croton lechleri*), a tree native to northwestern South America that yields a red resin known as dragon's blood. The plant-derived drug regulates intestinal water and prevents dehydration, thus providing a treatment for AIDS- and HIV-related diarrhea. In January 2013, the FDA approved Shaman's crofelemer, under the trade name Fulyzaq, for this purpose. The drug was also approved for treating infectious diarrhea in children, a leading cause of childhood death in some countries. A different study investigated the use of *C. lechleri* in comparison with taxol and vinblastine to control melanoma cancer cells. Researchers found that the plant medicine inhibited cancer cell proliferation, thus supporting the traditional use of the sap as an anticancer agent (Montopoli et al. 2012).

An ethics of reciprocity contrasts starkly with practices of biopiracy, defined as forms of bioprospecting that involve exploitation of indigenous knowledge by commercial entities. Geographer Daniel Robinson regards biopiracy as "the appropriation of biological resources and associated knowledge, particularly from the most biodiverse developing countries and from farmers, indigenous peoples and local communities" (Robinson 2010, 1). Key accounts of biopiracy include Vandana Shiva's *Biopiracy: The Plunder of Nature and Knowledge* (1997) and Darrell Posey's work in ethnobiology (Posey 2002), both of which argue for ethical standards governing the commodification of traditional knowledge, including of medicinal plants. However, while critics of biopiracy advocate stronger rights for indigenous groups and the protection of the intellectual property related to their traditional ecological knowledge, an ethics of reciprocity with non-human species has not figured into their analyses. As such, conceptualizations of biopiracy are strongly human-centered, without the multispecies focus of posthumanist thinking.

In this section, we have seen both traditional and commodity-based examples of reciprocity, as well as the broad importance of giving back to plants and people for the good they provide. Indeed the act of giving back to the vegetal world can take a multitude of forms, which need to be identified, conceptualized, and designed according to the plant species, human communities, indigenous traditions, and biological habitats involved. However, by committing to reciprocity ethics, researchers, activists, and community members can ensure not only the longevity of plants species in their native habitats but their flourishing and wellbeing as subjects-of-a-life. An ethics of reciprocity can manifest as the intimate act of returning parts of the flower to the earth or the grander act of establishing ecological reserves to protect plant species and their wild habitats. Deciding which forms of reciprocity to put into practice depends on a range of factors (indigenous, cultural, social, ecological, botanical) that ensure the ethical purpose of these acts.

CONCLUSION: INDIGENEITY AND ETHNOBOTANICAL RECIPROCITY

The previous section briefly addressed the question, "what is 'cancer' in traditional indigenous knowledge systems, and what causes it?" The Navajo conceptualization of cancer reflects a close association with the ecology of south-western North America. As a primary causal agent, lightning is a form of radiation, which is furthermore a shooting phenomena. Arrows, lightning, and radiation figure into traditional Navajo cancer understandings in which modern causal agents (such as radiation) are integrated into an indigenous knowledge framework. The previous section also asked, "how does a traditional perspective of cancer influence the prescription or administration of botanical remedies?" A notable example from the Navajo study is the Pus-eater ceremony, and the herbal agents associated with it, for treating cancer as a condition of sores, pus, and decay. Moreover, in the traditional medical systems of China and India, the treatment of cancer involves treating the entire body, not solely the diseased organ. The efficacy of traditional tonics – *Oplopanax horridus*, *Withania somnifera*, *Panax ginseng*, and *Ganoderma lucidum* – in treating cancer suggests that health is a condition of the whole body, not only its parts. Whereas allopathy focuses a diagnosis on one organ or one bodily region, the ethnobotanical medicine of ancient China and India regards disease as an affliction of the entire organism, and thus prescribes tonics to build immune reserves and foster the body's systemic integrity.

The global prevalence of cancer indicates a basis in the byproducts of technological societies: "Cancer is evidently a much more important disease in modern America than it was in native America" (Moerman 1998). By contrast, the majority of diseases and disorders of traditional cultures stemmed from nutritional deficiency, injury, overexertion, and exposure to climatic extremes. Plants such as *Taxus* traditionally were used for rheumatism, colds, and lung disease. However, with the isolation of paclitaxel (taxol) from the Pacific yew tree (*Taxus brevifolia*), the same species has yielded a useful cancer drug, as reported in the WHO Model Lists of Essential Medicines and confirmed by the National Cancer Institute (Moerman 1998, 13). The Pacific Northwest Coast Tsimshian people have utilized *Taxus brevifolia* as a cancer treatment at the extreme north end of the range of the species. Although it is problematic to impose conventional medical ideas on traditional ethno-medical systems, the common factor of plants as sources of therapeutic agents brings both systems into dialogue. Ethnobotanists identify plants with anticancer properties by understanding traditional knowledge and searching for species that are known to be effective against conditions associated with cancer, such as inflammation. Plants used traditionally to treat "cancer" symptoms are identified and adopted by allopathic medicine after biochemical analysis, as the examples of crofelemer and taxol demonstrate. Moreover, as indicated by the use of African periwinkle as a leukemia treatment, the anticancer properties of a species are identified when researchers investigate seemingly unrelated potential applications of the plant in medicine.

However, as I have argued, the value of reciprocity is often left out of these ethno-medical processes where exchanges between plants, indigenous people, and the allopathic medical paradigm are focused on use-value: how effectively a species, such as dragon's blood, can be converted to an FDA-approved medicine or commodified through another means with no address to the plant as a subject-of-a-life. I refute Becker's purely virtue-theoretic approach to reciprocity ethics over an act-morality approach because ethnobotanical treatments for cancer are contingent on the wellbeing of plant lives in order to be effective agents for the well-being of human lives. As habitat loss and climate change threaten vegetal and human lives alike, there is no place for a virtue-theoric approach to reciprocity. Researchers estimate that the loss of global vascular plant diversity between 1995 and 2050 will be 25 percent. The most impacted ecosystems will be tropical woodlands and forests, savannahs, and shrublands. During the fifty year period between 2000–2050, land use disruption will contribute more to species loss rates and patterns than climate change (Van Vuuren, Sala, and Pereira 2006). We cannot wait for "the sense of obligation […] to appear to us […] in retrospect," as Becker would have it. A sense of obligation to medicinal plants should appear to us as foresight, as thinking about our duties to the plant world as part of the fabric of our exchanges with them. An ethics of reciprocity that is both deontological and prudential, in Callicott's terms, regards the land as a community of beings, not the least of which are the plants that provide humanity with the medicines of the past and future. The multiple forms that reciprocal exchanges can take should be the subject of further philosophical and ethnobotanical inquiry into posthuman plants.

PART II

Heritage & Digitality

Johnson Papyrus (5th Century AD). Fragment of an illustrated herbal

Natural Heritage Conservation and Eco-Digital Poiesis: A Western Australian Example

INTRODUCTION

A city of biodiversity, Perth, Western Australia, faces significant environmental challenges. As species and habitats vanish, so too can their biocultural heritage, including their medicinal value (see chapter 2). To address biological and cultural decline, FloraCultures is a digital conservation initiative that uses archival, ethnographic and design approaches to conserve and promote Perth's "botanical heritage." This chapter examines the project's conceptual foundations in terms of nature/culture, tangible/intangible, and thinking/making dualisms, as well as some of the practical strategies used to address these dualisms. To articulate biocultural heritage, I have had to rethink categorical oppositions through ecopoiesis – the making of interactive digital objects as informed by ecological discourses. The repository being developed will incorporate cultural materials (texts, visual art, interview recordings, music, and video) not conventionally associated with environmental conservation. Key community-building approaches, such as focus groups and crowdsourcing, discussed later in the chapter, provide digitally-based interventions into biocultural heritage loss that reflect the ecopoietic basis of FloraCultures.

The South-West of WA, including the Perth metropolitan area, is an internationally recognized biodiversity "hotspot" (Breeden and Breeden 2010) (also, see chapters 1 and 4). The region supports an endemic range of floral, faunal, and fungal species – many of which are seriously threatened or face extinction through climate change and rapid urbanization (Ryan 2014, 49–59). Situated in Perth, the FloraCultures project confronts such urgent realities of regional conservation, but through the ethos of critical heritage studies (Smith 2006) and the methods of digital artefact-making. The project will result in an online archive for conserving the cultural heritage of Perth's flora (www.FloraCultures.org.au). Outlining a theory and practice of botanical heritage conservation has necessitated the rethinking of conceptual distinctions and the rethinking of the vegetal. "Botanical heritage" is intrinsically biocultural and involves the dynamic interplay between people and plants in a place over time as expressed in works of art, cultural artefacts, historical perceptions, and popular values, beliefs, and attitudes (Ryan 2014, 49). However, defined as a biological form of heritage – as a "genetic storehouse" or "natural resource" – botanical heritage can be constructed through quantitative practices of conservation and positivist modes of knowledge-making.

In Western Australia, the relationship between indigenous plants, classificatory data, and digital technology is exemplified by the online tool FloraBase – the scientific analogue of FloraCultures (The Western Australian Herbarium 2014). To develop a repository of botanical heritage, I have needed to consider three persistent binaries critically and propose ways to address them practically. These binaries are nature/culture, tangible/intangible, and thinking/making. This chapter outlines the manner in which I have navigated the interdisciplinary theoretical terrain of FloraCultures and concludes, on a hopeful note, that "ecological poiesis" (or "ecopoiesis") offers a mediating space of dialogue and creativity. The digital manifestation of ecopoiesis is the artefact itself – an open-access online repository of multimedia material. In the long-term, digital interventions into botanical heritage conservation through FloraCultures require sustained interactions with user-contributors through focus groups and the crowdsourcing of archival content. The ideal end result will be the promotion of the heritage of Perth's flora, above and beyond its scientific value or biological composition (*i.e.*, as genes, tissues, species, or ecological communities).

BOTANICAL HERITAGE: FROM NATURAL AND CULTURAL TO BIOCULTURAL

FloraCultures is a pilot project (2013–14) I developed in collaboration with Kings Park and Botanic Garden in Perth with funding through a seed grant from Edith Cowan University. With support from a team of digital designers and archival researchers, I have been primarily responsible for the project's conceptualization, design and implementation. FloraCultures is a digitally-mediated biocultural conservation initiative. The project centralizes the interrelationships between categories of heritage through a focus on a cross-section of indigenous plants identified, in consultation with Kings Park staff, as having high heritage value (Ryan 2014, chapter 4). The aim is to generate an integrative framework for documenting and conserving the botanical heritage of about fifty species of the Kings Park bushland.

The FloraCultures methodology uses traditional archival, oral history, and digital design approaches. The initiative is based on the idea that natural heritage is cultural and biological. An activist ethos underlies the project; an appreciation of heritage and its different forms goes hand-in-hand with the protection of living plants in their habitats.To this effect, the online repository will showcase a broadly conceptualized suite of heritage content – including interviews with conservationists alongside works of cultural interest that derive from (or offer a perspective on) the flora of the city circa 1827 when the Swan River Colony was founded by British settlers. The participatory web resource will be of interest to seasonal tourists, amateur naturalists, botanical artists, natural history writers, heritage consultants, and environmental conservationists. FloraCultures reflects the belief that an appreciation of biodiversity for its cultural, social, historical, artistic, and literary value helps to sustain environmental conservation on the ground (Ryan 2014, chapter 5).

The integrative model broadens the practices of conventional heritage conservation work, particularly in relation to the nature/culture binary. As one of the most tenacious aspects of thinking, being, and making, the nature/culture formation remains a theoretical difficulty recognized by early cultural studies scholars (Williams 1982) and negotiated by environmental humanities and ecocultural researchers (Giblett 2011). In the discourse of natural science, culture tends to be regarded as a heritage of nature, whereas nature is constructed socio-culturally in the humanities and social sciences (Olwig 2006). In response to these contexts, the heritage theory and practice of FloraCultures attempts to deconstruct and navigate the limiting doctrine of separation between nature, culture, conservation, and heritage. However, nature/culture dualism is embedded in practices of heritage conservation. Natural heritage tends to be bifurcated from cultural heritage, the former overly narrowed through an emphasis on biological materialism (saving plants, soils, ecologies) and cultural artefacts (saving objects of museological importance made from plants).

An inclusive view of heritage has precedents in the field of heritage studies where the nature/culture distinction has been increasingly scrutinized and recast. Although not necessarily considered in terms of plants, this bifurcation has been critiqued by scholars of biocultural heritage (Harmon 2013, Papayannis and Howard 2013, Vidal 2011), digital heritage (Cameron and Kenderdine 2007), heritage, globalization, and the environmental crisis (Long and Smith 2010), and natural heritage (Convery and Davis forthcoming, Dorfman 2011, Olwig 2006). As informed by these studies, my approach to heritage – as situated between nature and culture – requires a practice of working across ingrained typologies, both in theory and application. Conceptual reflexivity ("nature-culture" rather than "nature/culture") decompartmentalizes the binaries and brings the sciences, arts, and humanities into transdisciplinary dialogue (Ryan 2012a, chapter 1). The simple yet potent assertion that "nature is a cultural category" (Giblett 2011, 15) underpins a biocultural conceptualization of heritage, leading to a more inclusive practice of conservation, both in the field and in the digital domain.

Researchers on "cultural landscapes" assert that "nature is an inextricable part of culture" (Papayannis and Howard 2013, ix). Inextricability is evident, for example, in the designations of cultural landscapes within the UNESCO World Heritage Convention and the European Landscape Convention (or the Florence Convention). Papayannis and Howard describe the "double impact" of regarding nature as cultural heritage. The first involves the close coupling of the natural world and human culture, broadening the premises of heritage conservation and expanding the range of what content ought to be included in archives. The second leads to a biocultural ethics that calls attention to the natural "dividends" passed through generations and forming the basis of cultural inheritance (Papayannis and Howard 2013, xi). Heritage practice is ecological in character; the preservation of nature as cultural inheritance involves the safeguarding of the processes that have given rise to diversity in all forms (Harmon 2001, 64–66).

Echoing notions of cultural landscapes and inheritance, other scholars point to "biocultural diversity" conservation as an integration of heritage methods (Harmon 2013, 77). For example, the project "Endangerment and its Consequences" of the Max Planck Institute for the History of Science examines "the blurring of boundaries between 'nature' and 'culture' and the emergence of biocultural diversity both as an intrinsically endangered phenomenon and as the goal of scientific and conservation projects" (Vidal 2011). In the 1990s, the idea of biocultural diversity gained traction as anthropologists and linguists identified the overlays between biological and linguistic loss, particularly between the extinction of plant and animals and the decline of endemic languages (Maffi 2008). The biocultural concept centers on correspondences between biological (*e.g.* genes, populations, species, ecosystems) and cultural diversity (*e.g.* linguistics and ethnobotanical knowledge) (Maffi 2001). Jonathan Loh and David Harmon theorize biocultural diversity as "the total variety exhibited by the world's natural and cultural systems" (Loh and Harmon 2005, 231). They further explain the concept as:

> The sum total of the world's differences, no matter what their origin. It includes biological diversity at all its levels, from genes to populations to species to ecosystems; cultural diversity in all its manifestations (including linguistic diversity), ranging from individual ideas to entire cultures; and, importantly, the interactions among all these [...] Conceptually, biocultural diversity   bridges   the   divide   between

disciplines in the social sciences [and the humanities I will add] that focus on human creativity and behavior, and those in the natural sciences that focus on the evolutionary fecundity of the non-human world. (Loh and Harmon 2005, 231–232)

For these authors, the outcome of a biocultural approach to heritage conservation is a "more integrated view of the patterns that characterize life on Earth" (Loh and Harmon 2005, 232). Moreover, the study of biocultural diversity spans theory, practice, politics, and ethics (Maffi 2005). Its ethical ramifications inflect the moral dimensions of cultural inheritance previously discussed.

   Drawing from environmental ethicists, Harmon asserts that the maintenance of the world's biocultural heritage should become a moral obligation. A regionally-based biocultural ethics would be shared among diverse parties, including nature conservationists, social scientists, and cultural archivists, as a basis of an integrated heritage framework. As a regional approach to biocultural ethics, Harmon (2013, 78) proposes the recognition of "biocultural hotspots" to augment the scientifically-based designation, "biodiversity hotspots" (Harmon 2013, 78). Biodiversity hotspots are "areas that hold exceptionally high levels of the planet's endemic plant and terrestrial vertebrate species and which also are losing large percentages of their natural habitat" (Harmon 2013, 78–79). The approach to heritage conservation in FloraCultures begins from this premise. The South-West region is not only an epicenter of biodiversity but also of biocultural diversity – cultural heritage is part and parcel of the biological heritage equation. Additionally, the plurality of botanical heritage (*i.e.* "cultures" rather than "culture") is vital, and spans traditional Aboriginal Australian knowledge of plants, colonial-era European writings and artworks, and contemporary immigrant perceptions of local Perth-area flora. For FloraCultures, cultural plurality also signifies the bringing together of the "two cultures" of the humanities and sciences in the conceptualization and production of a repository. However, while much has been invested in conserving the biodiversity of the South-West in which metropolitan Perth is situated (*i.e.* plants and their environments), work remains to be done to ensure that the region's biocultural diversity stays intact, alive, and accessible to current and future audiences.

THE DIGITAL REPOSITORY: FROM TANGIBLE AND INTANGIBLE TO INTEGRATIVE HERITAGE

Biocultural heritage integrates the natural and cultural; a digital repository becomes a site for reconfiguring these distinctions. As such, it is necessary to consider the relationship between the artefact itself and the conceptual work it facilitates. Digital media techniques have been applied to the conservation of cultural heritage (for example, MacDonald 2006). Moreover, the rise of digital technologies and new media in heritage conservation has been described extensively for its participatory potential (for example, Parry 2010). However, less has been published on the application of digital creativity to natural heritage protection and promotion (Brown 2007, Maffi and Woodley 2010). I argue that this discrepancy, in part, is a result of the nature/culture dualism, in which the onus of natural heritage protection falls to conservation science and allied scientific disciplines. Articulating a biocultural form of heritage for Perth – and attempting to conserve it through a conceptual framework (theory) and an online repository (practice) – has required consideration of the role of archival instruments in the digital era. Rather than static tools of preservation – the virtual equivalents of dusty archives, contained in a physical location and visited by specialist researchers – digital repositories can become community-engaged spaces of creative production, building relationships between users, participants, and conservators. The dynamic, interactive, and participatory possibilities of digital repositories are compatible with the genre-blending of heritage content in FloraCultures, in terms of nature and culture, as well as "tangible" and "intangible" forms (Dorfman 2011).

Indeed, botanical heritage (as biocultural) can be tangible, intangible, or both. Yet, like the nature/culture binary, these categorical distinctions risk slighting the interconnections between forms of heritage. The field of critical heritage studies interrogates these dualisms through actual practices of heritage conservation. For example, Laurajane Smith (2006) critiques the assumptions of authorized heritage discourses (AHD) in terms of what constitutes heritage (typically material artefacts and places, in her view) and the exclusion of the public from direct involvement in conservation processes. Smith's argument can be applied to botanical heritage. On the one hand, tangible botanical heritage (TBH) includes "materialized forms of cultural expression" (Lixinski 2013, 7) involving plants, for example, as architectural works constructed from local timber, items of clothing woven from plant fibers, or artisanal creations using plant dyes, flowers, or seeds. On the other, intangible botanical heritage (IBH) can be theorized as either dependent on or independent of tangible heritage. As dependent, intangible botanical heritage encompasses "the processes, skills, and beliefs leading to the creation of tangible works" (Lixinski 2013, 8). As independent, intangible botanical heritage refers to the memories, stories, songs, dances, ceremonies, and other knowledge forms involving plants that do not necessarily have fixed material reference points (Ryan 2012, see chapter 8 on botanical memory). However, as FloraCultures indicates, all intangible botanical heritage is dependent on, referential to, or triggered by material artefacts to some extent. Human memories of nature are catalyzed or deepened by direct reference to living things, meaningful objects, or important places. For instance, songs might necessitate real instruments made from tree bark or ceremonies might center on the use of an aromatic resin from a local species. In other words, tangible botanical heritage and intangible botanical heritage are inextricably related in theory and practice; the division between tangible and intangible is a false binary, especially considering the role of living flora and plant-based objects in prompting intangible heritage (memories, stories, ceremonial knowledge, etc.).

In the Perth context, tangible botanical heritage is scattered across a number of physical locations, such as the archives of the WA Museum and small private collections. Similarly dispersed, the intangible botanical heritage of the area is evident in recorded oral histories with Aboriginal Australian and Anglo-European interviewees, as well as extant textual, audio, and video material. The FloraCultures repository is designed to conserve both forms of heritage, that is, to make visible the complementarity between the tangible and intangible in terms of plants. Indeed, the multifaceted potential of a digital repository offers a space which can manifest past, present, and future dynamics between heritage forms. One of the roles of an online repository such as FloraCultures is as a collection of digitized cultural artifacts – including texts and visual artworks – designed to give the user an impression of the region's botanical heritage. However, the repository also serves as a signifier of tangible botanical heritage, involving photographic and written documentation of plant-based artifacts existing in a physical space somewhere else, such as the private collections of a botanical artist. Yet another identity of a repository is as a research tool or a taxonomy of the biocultural world – a database – that allows users to search for and locate content systematically. In the digital era, a repository is additionally an interactive, educational, or promotional platform, making possible the crowdsourcing of heritage material, including family-based memories and community-shared stories. In the latter two senses, the repository is a creative space for artists, writers, activists, researchers, and concerned citizens or, in Boris Groys' terms, a "living machine" capable of change, adaptation, and decline.

For Groys, the boundaries of a biocultural repository are fluid and changing. The determination of what is significant and, therefore, to be included in a repository (what he terms the "New") and what is irrelevant and, therefore, to be excluded (the "Old" or "noncollected reality") reflects the dynamic contexts in which heritage is always produced and situated (Groys 2012, 1–2). More than a representational system or posterior instrument, the repository provides the basis for historical, cultural, and artistic creation, as "a machine for the production of memories, a machine that fabricates history out of the material of noncollected reality" (Groys 2012, 3). A biocultural repository is not designed as a static artifact available to a privileged few. Digital preservation and access are dynamic processes, engaging networks of users, creators, and conservators and involving citizen archivists and community members (Prelinger 2009). As a long-term resource, an online repository facilitates "cultural production" (rather than cultural preservation only) from preserved content, reflecting the Creative Commons notion of "free culture," which limits the reach of restrictions on creative (re)use of material (Lessig 2004).

The theory and practice of the FloraCultures repository are designed to move seamlessly across forms, in recognition of the intrinsic links between these typologies. As a digital artefact – rather than a physical space where heritage objects can be preserved and displayed – FloraCultures is poised to foreground tangible-intangible connections. For this reason, I have been conducting oral histories with individuals with cultural knowledge of the plants of Kings Park. Recollections of certain species lead to memories of botanically rich places outside of the park that have been drastically impacted by the recent clearing of the bush. For example, Kim Fletcher, a Kings Park Volunteer Guide for over ten years, reflects on his life-long interest in orchids, fostered by childhood excursions throughout the Armadale area in the southern suburbs: "Underneath the marri (*Corymbia calophylla*) I remember these pink enamel orchids, not the purple ones, much bigger and popping up through all the leaf litter. So glossy looking. They were beautiful" (see chapter 1 for more about this interview). Kim's memories meander between Perth and Armadale – his recollections not circumscribed by the physical limits of Kings Park. Moreover, in an interview, the memories and stories of botanical artist Nalda Searles are continually elicited by touching, smelling, and pointing tothe plant-based creations surrounding us in her home, demonstrating the correlation between tangible heritage (*e.g.* botanical works of art made from balga (*Xanthorrhoea preissii*) leaves) and intangible heritage (*e.g.* memories of collecting the leaves and knowledge of their tactile properties).

ECOPOIESIS: FROM THINKING AND MAKING TO ENVIRONMENTAL PRAXIS

As a digital object with a conceptual underpinning, the repository underscores the dynamics between thinking (theorizing, critiquing, analyzing) and making (doing, producing, materializing). However, a theme in the literature of the digital humanities posits critique as either separate from making or a relic of traditional humanistic inquiry and, in particular, the practice of textual analysis. For example, although he is willing to admit critical theorists into the field, Stephen Ramsay (2011a) asserts that "personally I think Digital Humanities is about building things [...] if you are not making anything, you are not [...] a digital humanist." Moreover, for Ramsay, the digital humanities involve "moving from reading and critiquing to building and making," even at the risk of being "undertheorised" (Ramsay 2011b). Just as natural/cultural and tangible/intangible typologies limit eco-digital humanities research, so too does the thinking/making binary reinforced by some theorists. In contrast, FloraCultures involves thinking-making reflexivity in which each iteratively sculpts the other.

Practice-led research offers a basis for rethinking theory/practice in the digital humanities. In his "A Manifesto for Performative Research," Brad Haseman describes practice-led as experiential research resulting in new forms of performance or exhibition, or interactive digital objects, such as games. What he calls "performative research" (as a third research paradigm after quantitative and qualitative paradigms) is suited to user-led projects and end-user research (Haseman 2006, 9). In particular, the approach engages "the processes of trialing and prototyping [...] in the development of research applications in online education, virtual heritage, creative retail, cultural tourism and business-to-consumer applications" (Haseman 2006, 9). Practice-led research is "initiated" and "carried out through" practice (Gray 1996). Exegetical commentary in practice-led projects explicates the central role of thinking-making reflexivity, in which the object (*e.g.* a text, performance, composition, or repository) is more than situated in, but is also shaped by, critical discourse.

In conjunction with the practice-led model, a concept which helps to resolve the theory/practice opposition is ecopoiesis – borrowed from ecocriticism and applied here to eco-digital productions (for example, Mules 2014, Rigby 2004). I define "digital ecopoiesis" as the making of interactive digital objects, shaped by ecological discourses, such as ecofeminism, bioregionalism, and sustainability. Poiesis is "making" or "producing," whereas praxis is "doing" or "acting" (Mules 2014, 21). In a phenomenological sense, "poiesis" means "bringing forth" – the dynamic capacity of things (animate and inanimate) to change, adapt, or decline.

Moreover, physis, as the material becoming of nature (*e.g.* seeds bursting, flowers opening), is poietic bringing forth. Mules (2014, 22) argues that an object (biological or digital) is the poiesis it manifests and that the concept "identifies the being of things in their becoming other: in their creative, shaped and connected possibilities." The poiesis exhibited by a botanical artwork is related to the plant's material becoming, leading to a condition of "co-becoming other" between the creative work and the living species, and between species within an ecosystem (Mules 2014, 22). For Kate Rigby (2004, 440), ecopoiesis involves an "enhanced understanding of the natural world" and "technologies that are more compatible with its continued flourishing." Her interpretation of ecopoiesis encompasses praxis in relation to digitality: "Poiesis extends ultimately to a whole way of life. As such it is itself a form of praxis" (Rigby 2004, 430).

Ecopoiesis heralds the possibilities between environmental thinking and making in digital contexts. Here it should be noted that not all digital theorists subscribe to the thinking-making dichotomy. For these scholars, the making of objects enables ecological concepts to concretize and, conversely, the making of concepts allows eco-digital objects to come into being. Johanna Drucker (2009) argues that "making things, as a thinking practice, is not only formative but transformative." "Iterative conceptualisation" refers to "the means by which intellectual work takes shape (literally and metaphorically)" (Drucker 2009, 31). Moreover, "tinkering" is an iterative form of experimentation with digital objects – a process of shifting between thinking and making, rather than from the former to the latter, as Ramsay would have it (Jones 2013, 179). At the center of FloraCultures is ecopoiesis – mediating nature/culture, tangible/intangible, and thinking/making dualism and, therefore, creating pathways for digital interventions into South-West WA biocultural diversity loss. The outcome is a praxis (in Rigby's joined sense of poiesis and praxis, thinking and making) of botanical heritage protection.

FLORACULTURES: FROM COLLECTING AND ARCHIVING TO CROWDSOURCING

FloraCultures uses community-building processes to safeguard and procure botanical works of heritage value. In addition to conserving digitized versions of "objects" (*e.g.* floral illustrations or botanical poetry), the repository aims to crowdsource extant material while fostering the creation of new works by local artists and researchers. The repository operates across temporal levels – encompassing the more widely known historical records (*e.g.* the nineteenth-century writings of George Fletcher Moore), non-collected heritage content existing in smaller collections, and future works by botanical thinkers and makers. This multi-temporality is an ecopoiesis of heritage conservation; the digital repository is an object "in-the-making," in which archival material initiates the ongoing creation of biocultural heritage.

One of the project's community-building approaches is "design thinking," or "the methods and processes for investigating challenges, acquiring information, analyzing knowledge, and positioning solutions in the design and planning fields" (Plattner 2012, v). This "style of thinking" involves empathy, creativity, and reason in collaboration with potential users during different phases of a project (Plattner 2012, v). Informed by design thinking, in 2013 I conducted focus groups with end-users, including Kings Park, the WA Wildflower Society, Cockburn Wetlands Center, and an informal botanical artists collective (including writers). These parties represent botanical education, propagation, field conservation, and creative production. In response to a questionnaire, all Kings Park respondents agreed that the project offers a means for educating the public about biocultural heritage. More than 75 percent of respondents already use cultural content in educational tours or writings. However, 90 percent indicated an interest in accessing more Aboriginal knowledge of plants, as well as nineteenth and twentieth-century art and literature. In design terms, 85 percent agreed on the value of a repository search function using Aboriginal, common, and scientific names.

In conjunction with design thinking, crowdsourcing is a common dimension of participatory archival work. It is a problem-solving approach that enables an initiative to acquire content, services, or concepts through a "crowd" (*i.e.* a group or community) (Brabham 2013, 120–121). Eco-digital creativity emerges at the interface between the crowd and the institution.This approach enables expressions of botanical heritage to be captured. Although it lacks a physical archive, FloraCultures has attracted donations of bioculturally significant artifacts, such as sculptures and diaries, some of which can be digitized or, at least, digitally documented. Moreover, users will be invited to upload personal recollections or community knowledge of plants, thereby crowdsourcing intangible heritage. The result will be a web of stories about people, place, and plants, juxtaposed to images and written commentaries, prompting the input of other users. The ecopoietic web will lead to contributions from the botanical "crowd" of Perth, many of whom are accomplished botanists or artists.

Like the Kings Park discussions, focus groups with botanical artists centered on design possibilities, but with an interest in social media and online community development. In these sessions, I gathered input on the layout of the proposed repository but also encouraged "the crowd" to consider the content they already had (*e.g.* personal letters of colonial-era artists) or could create (*e.g.* visual artworks to be based on plants). Rather than a static device, the repository – well before its completion – has already become a "living machine" for past, present, and future biocultural heritage work.

CONCLUSION: THE CULTURAL SIGNIFICANCE OF FLORA

In developing FloraCultures, I have needed to reconsider the binaries of nature/culture, tangible/intangible, and thinking/making. The iterative process of decoupling and reformulating these linkages has happened alongside the work of biocultural conservation and the making of a digital repository from the ground up. The documentation of botanical heritage in all its expressions necessitates the intermingling of these conceptual formulations. This chapter has argued that ecopoiesis provides a middle path – a way of thinking about the eco-digital object as an agentic work-in-becoming that is always natural and cultural. It is via this middle path that the heritage value of Perth's flora, and the flora of other parts of the world, can be articulated and appreciated. The next chapter builds upon these themes, positing the posthuman plant as an ecodigital being.

CHAPTER 4

The Virtual and the Vegetal: Creating a "Living" Biocultural Heritage Archive through Digital Storytelling Approaches

INTRODUCTION

As the previous chapter articulated, FloraCultures is an online archive currently being developed in consultation with Kings Park and Botanic Garden in Perth, WA (see chapter 3). The archive will showcase the "botanical heritage" of indigenous plant species found in the extant bushland areas of Kings Park near the heart of the city. A selection of multimedia content (text, images, audio recordings, video interviews) and social media approaches (crowd-sourcing, interactivity, participatory media) will be brought together to highlight the cultural value of Perth's biocultural diversity. This chapter will further analyze FloraCultures in terms of Stuart Hall and Jacques Derrida's theories of "the living archive" in tandem with recent research into "digital storytelling" through new media. Derrida argues that the living archive is brought into existence through the dialectic between the death drive (Thanatos) and the conservation drive (Eros), and that an interdisciplinary field of "archiviology" is required to understand and develop archives in their broader cultural contexts. For Hall, the living archive is defined by heterodoxy as a participatory space consisting of a multitude of materials and in which public exchange can be fostered. I argue that a living archive in the digital era is brought to life through digital storytelling techniques that allow users to contribute to, participate in, and create their own stories as part of an ecology of the archive. In ecological terms, FloraCultures brings plant diversity – and the factors which impact it – to bear on the archive and the archivable. The conceptualization of the posthuman plant requires digital thinking.

The pilot project (2013–15) focuses on a selection of indigenous plant species found in the bushland areas of Kings Park and Botanic Garden in the capital city Perth (see chapter 3 for more detail). The aim is two-fold: to develop a concept of "botanical heritage" through a broad, interdisciplinary, and multimedia array of materials, including textual works, visual art, and oral histories; and to develop an online structure to preserve these digital (and digitized) artefacts that also allows the public to upload their own heritage content: memories, narratives, historical writings, visual artworks, and other artefacts related to Western Australian plants. The desired outcome of the FloraCultures archive is a dynamic digital environment weaving together extant and user-generated or user-sourced content through archival, ethnographic, and digital design strategies.

In this chapter, I will attempt to develop the conceptual foundation of FloraCultures further in relation to theories of "the living archive" (Derrida and Prenowitz 1995, Hall 2001, Refsland, Tuters, and Cooley 2007) and the exploration of "nonlinear narratives" through digital storytelling approaches (Alexander 2011, Lambert 2013, Sanderson 2009). In applying theories of the living archive to FloraCultures, I will consider, in detail, Jacques Derrida's Freudian theory of the archive presented in *Archive Fever* (1995) and Stuart Hall's articulation of heterodoxy as a core value in archival work. I conclude that, rather than a living archive, FloraCultures can be conceptualized as an "ecology of the archive" – a dynamic and interconnected (not a metaphorically living) eco-digital system that integrates diverse living communities of users (present and future) as well as heritage materials of disparate historical, botanical, and disciplinary provenance.

FLORACULTURES: A BIO-CULTURAL HERITAGE ARCHIVE

FloraCultures is presently being designed in consultation with Kings Park and Botanic Garden – a popular plant conservation, education, and tourism institution, located in Perth. In addition to a series of cultivated garden installations featuring plant communities from different parts of WA, Kings Park preserves extensive non-cultivated bushland areas consisting, for the most part, of indigenous plants (see Prologue). The pilot project is designed to call attention to and promote community engagement with the "botanical heritage" of Kings Park's bushland plants, including iconic Western Australian species such as kangaroo paws, donkey orchids, banksias, and marri trees (Ryan 2014, 49–59). The extant botanical diversity of Kings Park is representative of adjacent urban and suburban areas (including the Perth CBD, Northbridge area and western suburbs) where much of biodiversity has been fragmented or lost.

The approach to natural heritage conservation developed in FloraCultures reflects the ethos that biological and cultural heritage forms are interwoven – an interdependence expressed in the concept "biocultural diversity" posited in recent years by anthropologists, historians, and heritage scholars (see, for example, Maffi 2001). In theory and practice, the interdisciplinarity of botanical heritage – as both biological and cultural, that is, as biocultural – is an asset in terms of archival scope, inclusiveness, and prospective relevance to different users of the present and future. Circumscribed scientifically, botanical heritage would prioritize the protection of living plants in their natural habitats through the empirical methods of conservation science (*e.g.* seed propagation, habitat protection, weed control, or the introduction of pollinators). Although efficient strategies for conserving plants, these approaches tend to exclude the cultural, social, artistic, and intangible dimensions of human-plant interactions in a place over time (Ryan 2012a). Seeking a middle ground between disciplines – and more broadly between the sciences and the humanities – FloraCultures examines the complex intersections between cultural and biological heritage, where the decline of plants in the environment (*i.e.* living, growing organisms) affects the vitality of the cultural heritage involving those plants (*i.e.* paintings, poetry, music, memories) (see chapter 3).

Aiming for inclusiveness, plurality, and interactivity, FloraCultures combines multimedia content (text, images, audio recordings, video interviews) with social media approaches (crowd-sourcing, user-generated content) to underscore the importance of Perth's biodiversity. For as Derrida comments, "the question of the archive is not […] a question of the past" (Derrida and Prenowitz 1995, 27). Known as one of the world's most biodiverse urban areas, Perth and surrounding suburbs are rapidly losing biodiversity as a consequence of bushland clearing, plant diseases, and other factors. The continued loss of irreplaceable flora and fauna in Perth is a constant, looming reality for those working in conservation. While the building of a biocultural heritage archive will not necessarily protect actual biota in their habitats, it does offer a compelling way to rationalize their ongoing protection by appealing to cultural values and fostering community education. The digital archive features poetry, literary extracts, music, film clips, visual art, photography, historical documents, and oral histories with contemporary plant conservationists and artists – all focused on Kings Park bushland species and their importance to the city's heritage, identity, and wellbeing. It is hoped that FloraCultures, as a virtual repository of vegetal heritage, will offer a precedent for the conservation of Australian biocultural diversity – those manifestations of cultural heritage that strongly depend on the continued existence of plants, animals, fungi, rivers, mountains, and bioregions as a whole; and recognition of nature's right-to-exist (see chapter 2).

THE LIVING ARCHIVE: BETWEEN THANATOS AND EROS

A typical dictionary definition of the word "archive" (as a noun) is "a place or collection containing records, documents or other materials of historical interest"

(The Free Dictionary 2014). However, as scholars of the archive point out, the term has become a "loose signifier for a disparate set of concepts" (Manoff 2004, 10) – the ambiguous status of the term especially compounded by the introduction of digital technologies and social media approaches to archival practices in the last twenty years. Moreover, as a virtual "place" of both conservation and production – thus concerned simultaneously with the preservation of the past and the construction of the future, in Derrida's terms – a standard digital archive tends to comprise diverse forms of heritage material in digital format, including texts, images, and recordings. For example, the National Library of Australia's open-access Trove archive – a significant tool for researchers on Australian history, culture, and art – contains photos, music, video, maps, diaries, letters, and digitized newspapers. Most outstandingly, the NLA archivists have successfully implemented a crowd-sourced approach to historical documents, in which archive users help to transcribe digitized newspaper articles (National Library of Australia 2014).

Another digital archive of note and arguably one of the first to exist is UbuWeb, founded in 1996 by American poet Kenneth Goldsmith. UbuWeb is a volunteer-based, curated collection of "avant-garde" artworks, including sound art, visual works, film, and poetry (UbuWeb 2014). Particularly focusing on obscure, out-of-print, and limited-run works, the archive is replete with "the detritus and ephemera of great artists" (Goldsmith 2011). For example, the Andy Warhol Audio Archive contains audio interviews with the artist from between 1965 and 1987, as well as recordings of contemporary Canadian filmmaker David Cronenberg reflecting on Warhol's evolution as an artist.

But what does it mean for an archive, such as Trove or UbuWeb, to be "living" rather than simply retrospective, enduring or useful? Is the term "living" merely a hyperbolic flourish, or can an archive exhibit a peculiarly living form of agency within the larger cultural and ecological systems of which it is part? Indeed, the question of the living archive occupied Jacques Derrida in his treatise *Archive Fever*, which posits a Freudian theory of the archive (Derrida and Prenowitz 1995). Derrida's theorization of the archive draws largely from Freud's *Beyond the Pleasure Principle* (1961). According to Derrida, there are two opposing forces constituting the economy of the archive. The first is the death or nihilism drive associated with Thanatos, whereas the second is the archival or conservation drive linked to pleasure and Eros. Derrida describes the death drive as "anarchivic," "archiviolithic," or archive-destroying (Derrida and Prenowitz 1995, 14). He summarizes his Freudian interpretation of the archive in the following passage:

> Another economy is thus at work, the transaction between this death
> drive and the pleasure principle, between Thanatos and Eros, but also
> between the death drive and this seeming dual opposition of principles,
> of *arkhai*, for example the reality principle and the pleasure principle.
> (Derrida and Prenowitz 1995, 14)

The archive, thus, resists the reality of the death drive towards nihilism, entropy,
and loss. Like the biblical ark, the archive salvages the seeds of the past for the
fruition of the future despite the overwhelming archiviolithic floodwaters of
physical decay, technological obsolescence, cultural obscurity, social disregard,
and economic penury.

Furthermore, Derrida claims – in terms that recognize its "living" agency –
that the archive itself actively shapes history, memory, and the very nature of the
archivable. He describes the dynamic interplay between the archive and the
archivable as a process of "archivization" and claims that:

> … the technical structure of the *archiving* archive also determines the
> structure of the *archivable* content even in its very coming into existence
> and in its relationship to the future. The archivization produces as much
> as it records the event [italics in original] (Derrida and Prenowitz 1995,
> 17).

In other words, archival meaning is "codetermined by the structure that archives"
(Derrida and Prenowitz 1995, 18). Derrida's use of the verb "codetermine"
implies a dialectical relationship between the death drive and the conservation
drive, between Thanatos and Eros, between mortality and pleasure, between
anarchivic decay and the archivic future, or – to borrow philosopher Boris Groys'
terms – between non-collected reality (or the Old) and archivable reality (or the
New) (Groys 2012, 3). The archive is a liminal space emerging at the threshold of
these polarities. In light of such complexities, Derrida proposes the field of
"archiviology" as "a general and interdisciplinary science of the archive" (Derrida
and Prenowitz 1995, 26) – one informed by the expertise of historians, librarians,
artists, literary scholars, cultural theorists, heritage advocates, and computer
programmers.

For Stuart Hall, heterodoxy defines the living archive on all levels.
Influenced by the post-Structuralist thinking of Foucault, Hall argues that the
living archive is inherently a heterogeneous and dialogic place of intellectual
debate, public engagement, and social responsibility. The living archive is a
"never-completed project" that is "on-going, continuing, unfinished, open-ended"

(Hall 2001, 89), continually countering the "fantasy of completeness" (91). Yet the open-ended archive is not a chaotic construction or fortuitous event, but instead comes to life through an internal cohesion based on a kind of physiology – a set of ordering principles that determines the flow of non-collected reality into and out of the archive-as-system. More than a rote structure or mnemonic mechanism, the archive is first and foremost a "discursive formation" characterized by a "heterogeneity of topics and texts, of subjects and themes" (Hall 2001, 90). The texts composing the living archive are also heterogeneous, including personal stories, narratives, anecdotes, impressions, and biographies – in written, visual, aural, and mixed-media formats. Echoing Derrida, Hall (2001, 92) goes on to underscore that the living archive is never a static historical collection – concerned merely with the conservation of the past – but is a negotiated space, always in "active, dialogic relation to the questions which the present puts to the past."

Moreover, the practice of archiving (the structure and methodology), like the archivable (the content), is similarly diverse and dynamic, shifting between the private and public spheres:

> It includes those inert collections which have emerged, fortuitously, when odd individuals record or purchase works over time – works which may not be exhibited or accessible to anyone who is trying to do an archaeology of practice. That is the most buried, most inaccessible, most un-recoverable end of the archive. At the other end are the public spaces which have conscious policies of collection and selection, of display and access. (Hall 2001, 91)

Most importantly for Hall, the living archive must remain democratically open and accessible, requiring the energy of public involvement to foster "vitality of argument, debate and reinterpretation" (Hall 2001, 91). The issue of archival access is related to institutional contexts of funding and technology as well as to intellectual questions of aesthetics and interpretation – each factor not an independent strand but rather contingent on the nexus. Despite the intricacies between archival work and its broader social, cultural, economic, and ideological contexts, "it is extremely important that archives are committed to inclusiveness, since it is impossible to foretell what future practitioners, critics and historians will want to make of it" (Hall 2001, 92).

How do Derrida and Hall's conceptualizations of the living archive play out in the FloraCultures project? To begin with, heterogeneity is evident on different levels. The placing of equal importance on Aboriginal (Whadjuck Nyoongar, the first people of Perth), colonial Anglo-European settler, and recent immigrant content, foregrounds cultural plurality as an overarching value – FloraCultures. Diverse cultures, both pre- and post-colonial, sharing a geographical place over time have produced tangible and intangible "artifacts" (stories, texts, artworks, music) connected to Perth's indigenous flora. I propose that, through the pilot project, these artifacts in their collectivity come to express the botanical heritage of the city. Again, botanical heritage is more than exclusively a biological or material reality; it is always biocultural. Moreover, heterogeneity is visible in the temporality and multi-disciplinarity of the heritage content itself: visual art, textual works, oral histories, mixed-media expressions, from the pre-settlement years of the Swan River Colony (prior to 1829) to the 21st century narratives of conservationists involved with Kings Park's indigenous vegetation, for example, as educators or propagators. The textual heterogeneity spans historical accounts from the State's archives in which early naturalists recorded their impressions of Perth's eucalypts, alongside extracts from recent oral histories in which local people describe the emotional pain of witnessing declines in rare orchid populations after land has been cleared for development and habitat has been permanently lost.

In sum, the interplay between culture, ecology, content, and time underlies the project. The archive's conceptual basis pivots on Hall's heterodoxy – an essential principle that is enhanced by digital strategies, specifically public participation in the archive through social media (*i.e* crowd-sourcing or Facebook-style conversation threads about particular archived items). Following Hall's argument for the value of heterodoxy, I aver that the blend of heritage material contributes to FloraCultures as a living archive – not an inert collection of historical or botanical content accessed only by specialist researchers. To borrow again from Hall, FloraCultures shuns the "fantasy of completeness" by recognizing the value of community-sourced and user-created heritage. The living archive becomes a place of creative production – in the present tense – where forthcoming material is inspired and created from "inert collections," for instance, as artists respond to the archive's historical facets in designing works about the marginal status of plant species and their environments. What results is a call-and-response between creators and content – the archive as cultural stimulus, the content as a constellation of prompts – in which new manifestations of plant-based cultural (*i.e.* biocultural) heritage are formulated and then installed in the archive according to an open-ended, community-based rhythm that is largely out of the hands of the archivists.

Furthermore, the value of heterodoxy functions on another level – which I will explore more fully later in this chapter – but it is essentially ecological in character. Briefly, I will suggest here that the heterogeneity of the archive mirrors the heterogeneity of the plants themselves (*i.e.* biodiversity) and the different conditions to which they are subjected (both natural and anthropogenic). Biocultural heritage conservation becomes more imperative as the living plants (on which the heritage is based) begin to vanish from the urban landscape under the weight of habitat destruction and plant disease. Despite the relevance of his theory of the archive, Derrida overlooks the ecological context out of which the archive and the archivable may arise. As a living archive, FloraCultures is always a confluence of multiple realities beyond the archive-archivable dialectic: ecological, cultural, technological, political. Therefore, the struggle between Thanatos and Eros, between mortality and pleasure, which inspirits the living archive also imprints the ecological upon the archival.

DIGITAL NARRATIVES: TELLING THE STORIES OF PERTH'S PLANTS

The heterodoxy of FloraCultures as a living archive results in the possibility of nonlinear narratives, allowing users to track self-generated pathways through the heritage content. The web portal invites user-participants to engage with artworks, literature, historical accounts and oral histories to stimulate their personal memories of plants in relation to their own beliefs, experiences, and backgrounds. Once the pilot archive is finished in 2015, it will be possible for users to navigate routes through the heritage material, forging unique plant narratives through four categories: species names (Nyoongar, common, scientific), genre (literature, poetry, historical writing, art, photography, oral history), media form (texts, visual, audio, video), or time period (pre-colonial, colonial era, contemporary). However, users will also be able to contribute digital material to the archive, thereby enlarging the narrative possibilities of the structure and its content.

Above and beyond the heritage ideals outlined earlier in this chapter, FloraCultures endeavors to tell the stories of Perth's plants – through a diverse collection of works – in the belief that no single narrative can be told about the natural world. Just as the cultures underlying the project are plural so too are the stories that emerge; and the stories are intrinsically about the relationships between plants, people, and nature in all their complexities rather than isolated species of flora. In this regard, Stuart Hall's "fantasy of completeness" relates not only to the archive and the archivable but also to the narratives that materialize. Invariably the story of each plant is "on-going, continuing, unfinished, open-ended" (Hall 2001, 89), just like the archive itself; each story consists of a multiplicity of stories where new stories are borne out of the fruits of old narratives preserved in historical records. In other words, the genesis of contemporary plant narratives goes hand in hand with the preservation of stories past; to invoke Derrida again, "the question of the archive is not [...] a question of the past."

The concept of the living archive – as set out in particular by Derrida and Hall – is closely akin to emerging approaches to digital storytelling through new media. In the context of FloraCultures and other digital projects, such as Rivers of Emotion outlined later in this section, the relationship between the living archive (as the body) and digital storytelling (as the life blood) is an intimate and interdependent one. We know that storytelling is a vital component of oral traditions, found in many cultures of the past and present across the world. Indeed, storytelling is an adaptive practice that has responded over the millennia to available media forms – human speech, rocks, the ground, skin, wood, print-based books, electronic technologies, and, later, digital media (Refsland, Tuters, and Cooley 2007, 411). While the emergence of print-based writing pushed storytelling towards non-interactivity (*i.e.* reading a book from beginning to end without the possibility of remixing the narrative sequence with one's questions, provocations, interpretations, or knowledge), digital technologies are considered to promise more flexible and participatory modes of narrative. For example, the common online technique of hyperlinking encourages users to build specific navigational pathways through digital content, resulting in user-driven narratives not comparable to flipping through the pages of an old-fashioned book or using a work's index to locate material out of the narrative sequence of print.

But what is digital storytelling? Bryan Alexander defines the term simply as "telling stories with digital technologies" and cites a range of examples, such as a podcast about medieval history, a blog-based or mobile phone novel, and a story about trauma told through Facebook and incorporating text, images, and video (Alexander 2011, 3). Digital stories can be fictional and factual, "brief or epic, wrought from a single medium or sprawling across dozens" (Alexander 2011, 3). Often community-based, the practice encompasses animation, audio, and transmedia combining, for example, film, mobile-place-based components, and analogue activities based at a physical location. In slight contrast, Joe Lambert asserts that digital storytelling is not merely a narrative told with a computer or other digital media. It is principally a movement "dedicated to de-centering authority" (Lambert 2013, 37).

As a democratic practice, digital storytelling is marked by community engagement and an ethos of personal transformation that promotes the storyteller's agency. As such, it is a vital form of participatory media that can be seen through three lenses: the degree of collaboration between a facilitator and a storyteller; the role of literary voice and style in the narrative; and the form the story takes. Lambert's model of participatory media making is based on the premise – following Henry Jenkins and other scholars – that media consumption is inherently a creative act. His pyramidal scheme posits a taxonomy of participatory forms, beginning with "constructive consumption (surfing mass media)," "intermediated consumption (surfing the web)" and "constructed consumption in context (games, fan films, sampling)" and culminating in the higher degrees of participation he calls "co-constructed (artist-led media projects)," "facilitated (digital storytelling and photo narratives)" and "do it yourself" which appears to be a highly independent hybrid category (Lambert 2013, 39). For Lambert, digital storytelling entails collaboration between a facilitator and the storyteller. In similar terms, other new media scholars argue that digital stories "radically alter the familiar triad of author-text-reader and in the process produce new kinds of narrative. In the digital realm, authorship is dispersed, collaborative and unstable" (Friedlander 2008, 179).

The nonlinear narratives of some digital storytelling projects give users access to events previously inconceivable and unimaginable. An example of environmentally focused digital storytelling project is Mannahatta, which offers a glimpse into New York's now heavily urbanized Manhattan Island in the early 1600s as the island was when European explorers first charted the Hudson River (Sanderson 2009). Based on extensive research into the island's ecology and the activities of Native American communities, such as setting fire to habitats to promote the abundance of desirable animals and plants, Mannahatta allows users to "peel back" the layers of New York's contemporary cosmopolitan identity to tell the stories of the city's original flora, fauna, landforms, wetlands, and human inhabitants.

Scholars of digital storytelling argue that projects such as Mannahatta are based on a "decentralised vision of virtual heritage" in which a multitude of voices (human and nonhuman) come together to tell different stories in and about a shared place (Refsland, Tuters, and Cooley 2007, 413). Although Mannahatta lacks a strong user-based ontology, the project excels in bridging multiple time scales, both geological and cultural, entailing different human interactions with the original New York landscape. However, the assertion that "virtual heritage could greatly improve its efficacy by developing user-centered and dynamic systems for nonlinear storytelling" (Refsland, Tuters, and Cooley 2007, 415) is better actualized in and more central to the aims of the Rivers of Emotion project, focusing on a crowd-sourced approach to the cultural value of the rivers of the Perth region.

Rivers of Emotion offers a noteworthy model of a user-centric space for nonlinear storytelling through the interplay of textual, visual, and aural material. It also provides an exemplar of interactivity for the trajectory of the FloraCultures pilot archive. Rivers of Emotion is "an emotional history of the Derbarl Yerrigan and Djarlgarro Beelier / the Swan and Canning Rivers" of the Perth region (Rivers of Emotion 2012). The web portal is structured into the categories "Riverscenes," "Riversights," "Riversounds," and "Riverstories," and is designed to make possible the public exchange of experiences, feelings, and memories about these two major Perth rivers and the aquatic systems of which they are part. Similar to the FloraCultures pilot archive, the project's focus is multi-sensorial, featuring oral and written content of "soundscapes, landscapes, visual, aural and emotionscapes" (Rivers of Emotion 2012). For example, within "Riverscenes," an entry from a contributor named "G. Pickering" from 19 January 2013 revolves around an image of a jellyfish accompanied by the caption "the light dances as the jellyfish journey through my frame...mesmerising, meditative, illuminated, illuminating – they move me." Emotional and artistic responses to nature are valued, fostered, and preserved in digital format. Moreover, "Riversounds" includes recordings of the memories of local residents. Stan Parks speaks of swimming in the Swan River as a boy growing up in North Fremantle in the 1950s (Rivers of Emotion 2012). In complement to the extensive ethnographic information, the project also includes archival images of the rivers, such as digitized versions of key paintings from the Swan River Colony era circa 1829 and later.

As a means to promote community engagement with and deeper appreciation of Perth's flora, the FloraCultures archive will employ interactive digital storytelling approaches in an attempt to convey – and to elicit from the public – the multiple stories of each of the 48 species featured in the pilot project. As suggested previously, the digital story is a heterogeneous one – invoking again Hall's notion of heterodoxy – based on an extensive historical record and comprising the scientific accounts of professional botanists and skilled experts (recorded as interviews transcribed to a written form but also available as streaming audio and video data), alongside the narratives of community activists, botanical educators, and wildflower enthusiasts – both long-term residents and short-term visitors to the State. In the style of Rivers of Emotion, users of FloraCultures will be able to contribute their comments, impressions, and memories – as well as any plant-based heritage content (*i.e.* photos, diaries, recordings, or ephemera) – directly to the archive, enlarging the bounds of the collective story told by the community, past and present. User-generated content in the form of memories and other plant narratives can be either provoked by existing material (*e.g.* someone remembers a story told by their grandparents after viewing a painting in the archive) or created in response to existing content (*e.g.* someone writes a short story about a plant species puzzled over by Swan River Colony settlers and then contributes the work to the archive). I believe that the crowdsourcing of plant-based heritage material is a core feature of nonlinear storytelling in digital environments in which traditional and relatively static archival material (texts, images, recordings) intermixes with dynamic new media based expressions of heritage.

CASE STUDY: THE DIGITAL HERITAGE OF THE SHEOAK

In order to demonstrate the spectrum of content of concern to the FloraCultures heritage methodology, I offer a brief example of the western sheoak (*Allocasuarina* fraseriana) – a common species in the Kings Park bushland and an endemic Western Australian plant with an extensive cultural history. As paintings, poems, and commentaries, these cultural resonances are not wholly related to the digital (*i.e.* they have had a long analogue existence before being digitized); nevertheless they demonstrate the movement between digital and traditional print-based materials that is at the core of the project. The process of uncovering the sheoak narrative(s) begins by tracing the intriguing etymological roots of its scientific, common, and Aboriginal names. The genus *Allocasuarina* combines the Greek root allos for "other" and its former genus name, *Casuarina*, which derives from the Latin *casuarius* for "cassowary" from the likeness of the tree's branches to the bird's feathers.

As its Latinate genus suggests, sheoak is "another kind of casuarina," one of many casuarinas across Australia. Interestingly, since the western sheoak was renamed *Allocasuarina*, the former denomination "casuarina" is more generally used now as a preferred common name by field naturalists instead of "sheoak" – an example of the fluidity between categories and practices of naming the natural world over time. The species name *fraseriana* honors colonial botanist Charles Fraser (1788–1831) – the first head of Sydney Botanical Gardens and one of the "fathers" of Australian botany. In Nyoongar terms, sheoak is called *gulli* or *kwela* (Moore 1978, 31, 46). *Kwel* in Nyoongar means "name," so sheoak is regarded by Nyoongar elders as the tree of naming – a plant that holds the names of everything and everyone, living and deceased, animate and inanimate, and that utters those names through its "whispers" (Noel Nannup, pers. comm., January 12, 2014).

Sheoak's name, aural environment and timber are also the subjects of many disparate yet overlapping stories throughout Western Australian colonial history. For instance, in the late 1800s, Western Australian settler Janet Millett wrote of its signature sound known as "sheoak whispers:"

> ... a few weird she-oaks destitute of leaves, between whose fine countless twigs, doing duty for foliage, the air sighs in passing with the sound as of a distant railway train. (Millett 1980)

Other Australian writers, such as the 19th century poet Charles Harpur, also observed the phenomenon of sheoak whispers, although often in more melancholic terms than Millett. As another part of the naming story, sheoak gained part of its common name from the similarity of its grain to its English namesake: the oak. Indeed, the origin of the prefix "she" has been the subject of speculation amongst foresters and botanists, offering another controversy related to plant naming (Ryan 2012a).

No matter its nomenclatural derivation, sheoaks across Australia certainly would have reminded early Anglo-European settlers of the treasured native oaks back home: a durable and craftable timber. However, like many of Australia's gum trees, the sheoak could have been considered inferior to its northern counterparts. Sheoak was used in settlement years for shingles and yokes, but, since the early 20th century, the wood has been prized for crafts, cabinets and furniture. For example, an article in *The West Australian* (1933) discusses a "presidential chair" constructed of sheoak and presented as a gift to the Pharmaceutical Society of Great Britain (The West Australian 1933). One-hundred years prior to the presidential chair, on August 12, 1829 the Foundation of Perth ceremony took place during which Helen Dance, the wife of Captain Dance of the HMAS Sulfur, cut down a symbolic sheoak near the present site of the Perth Town Hall. This momentous chopping of the tree is depicted in George Pitt Morrison's painting "The Foundation of Perth" (1929).

As these different short examples indicate, the sheoak has a rich Nyoongar and colonial cultural history. However, the casuarina has also been a fascination of Western Australian poets writing today. For instance, John Mateer's poem "Casuarina, the Word" offers a nuanced reinterpretation of the age-old sheoak whispers and the likening of its foliage to a cassowary's feathers, worth quoting in full:

> The word is a gaol, a plot of land, a tree
> and a cassowary. When we heard,
> we didn't name that sound
> as a cry, a call or a song.
> Maybe a cow, a fox, a devil
> or a cassowary? Not an emu in the guise
> of a sizzling tree, nor Macassan
> eyes mistaking the she-oak's feathered
> branches for the wings of a cassowary. (Mateer 2010, 68)

Poetry, such as Mateer's – along with cultural and historical content from different time periods, as cited earlier in this section – can serve as prompts within the digital heritage archive, stimulating the creation of new images, writings, and recordings from users in relation to a plant species, its sensorial environment, cultural interpretation, and ecological value. Intriguing themes within the broader story – exemplified by sheoak whispers – demonstrate how archival content might prompt future cultural production (*e.g.* artwork in response to the theme), which can then returned to the archive in a reflexive cycle of botanical heritage creation and public appreciation into the future.

CONCLUSION: TOWARD AN ECOLOGY OF THE ARCHIVE

This chapter has focused on the interconnections between virtual and vegetal heritage through the example of the FloraCultures archive. Derrida's Freudian analysis of the archive at the cusp of the struggle between Thanatos and Eros and Hall's privileging of heterodoxy and incompleteness collectively seem to posit a theory of the living archive. I have suggested that digital storytelling is a core principle and practice of any archive deemed to be "living" in which there is a blurring of the distinction between the traditional archive (as relatively static, fixed, and material) and new media based archives (as potentially more participatory, open-ended, and indeterminate). The archive as a "loose signifier for a disparate set of concepts" (Manoff 2004, 10) is increasingly the reality in the digital era as emerging (and emerged) technologies – especially the participatory legacies of Web 2.0 – rapidly transform archival philosophies and practices (Refsland, Tuters, and Cooley 2007, 409). Projects such as FloraCultures and Rivers of Emotions suggest that the archive of new media provenance is a heterogeneous space in-becoming – one which challenges pre-existing concepts of what an archive should resemble. The brief example of the sheoak demonstrates the range of content to be included in FloraCultures and how extant material can invigorate the aims of botanical conservation and artistic creation.

But what precisely makes FloraCultures different from other multimedia archives, such as Trove, UbuWeb, and Rivers of Emotion? If we accept that all archives are "living" – a notion applied somewhat uncritically and sweepingly by both Derrida and Hall – then how is FloraCultures unique? I suggest that the diachronic and dialogic qualities of the project – as well as its medial heterodoxy – distinguish FloraCultures as the first archive of its kind to attempt to engage seriously with the lives of plants. In comparison, other related botanical heritage archives are narrowly focused on specific aspects of the botanical world – usually utilitarian in emphasis. For example, the University of Michigan's Native American Ethnobotany database allows online users to research the ethno-botanical aspects of North American plants as food, fiber, and medicine (University of Michigan 2014). However, the repository disregards the significance of those plants to the art, literature, and cultural identities of Indigenous peoples, Anglo-European colonists, and contemporary American societies.

Similarly, Kew's web resource Useful Plants and Fungi (formerly called Plant Cultures) focuses on the ethno-botanical importance of a range of plants as building materials, fibres, dyes, food, drink, fuel, and medicine (Kew Royal Botanic Gardens 2014). While the website underscores the inextricability of flora and human cultures globally, the master narrative that results is one of anthropocentric exploitation that marginalizes the agency of plants. Furthermore, these two repositories of heritage content offer the community no means to engage, participate, or contribute – thus constraining the possibility that botanical heritage is constituted in the past, present, and future by specialists, scientists, artists, writers, and the public alike. In other words, what these botanical heritage archives lack is an ecology of archival practice – a digital ecology that Derrida and Hall fail to articulate in their accounts of the archive.

Rather than an archive as an economy (a mechanistic, fiduciary metaphor) or a living thing (a biological, material figuration), in Derrida and Hall's terms, I prefer to conceptualize the archive as an ecology – a dynamic and interdependent "system" of animate and inanimate actants. If we accept that a natural heritage archive (or any archive for that matter) will never actually be a living thing but rather a referent for living things, then what is the use of this trope that risks diminishing the agency of real organic beings, such as plants? Nevertheless, it is productive to recognize that the agency of the archive depends on the agency of networks of past, present, and future user-contributors and the plants themselves. These networks are given shape by the form of the digital archive, the heritage content that becomes archivable and the changing ecological and cultural contexts which yield biocultural heritage in the first place. Between the archive and archivable – between entropy and conservation – is an ecological flux that makes a project such as FloraCultures an open-ended, ever-changing, and vastly interconnected system. The archival relationship between the virtual and vegetal is, therefore, determined by the environmental dialectic between Eros and Thanatos in urban Perth and anywhere else for that matter. In simpler terms, botanical heritage becomes ever-more jeopardized when healthy plant communities no longer exist in proximity to people, especially in terms of intangible cultural heritage (ICH) such as memories, impressions, and stories.

FloraCultures as an ecology recognizes that the decline of indigenous plants in everyday life in Perth or the conservation of those plants through the sustained efforts of human communities impacts the composition of the archive. Again invoking Derrida, the archive is the dialectic between the drive towards pleasure (Eros, conservation) and the drive towards entropy (Thanatos, destruction). The FloraCultures archive, thus, addresses the pathos of species loss in the urban environment of Perth through the digital conservation of botanical heritage (rather than through other means, such as field-based conservation). It should be stressed that the material referents for digital botanical heritage are always the living plants themselves, a codetermination between archive-archivable-ecology in which "ecology" is both natural and cultural (involving the complex relationships between plants, people, and the natural world over time).

This position should not be read as a form of "digital logocentrism" in which the user interactivity, medial heterodoxy, and diachronic flux of the digital archive supersede the relative fixity of the analogue archive. Instead, the digital archive as a participatory ecology – rather than an economy or living thing – brings into dialogue (but does not conflate) the ecological focus of the archive and the networks of living and non-living actants (plants, environments, people, digital technologies, digital, and analogue or print-based artefacts) upon which the archive is built. Subsequent research into FloraCultures and other botanical archives that involve digital structures should focus on how the media technologies of archivization set, modify, or legitimate forms of environmental memorialization. Such understandings of the digital also bring into focus the posthuman plant as an ecodigital being, a subject of particular relevance to the next part of this book.

PART III

Art & Vegetality

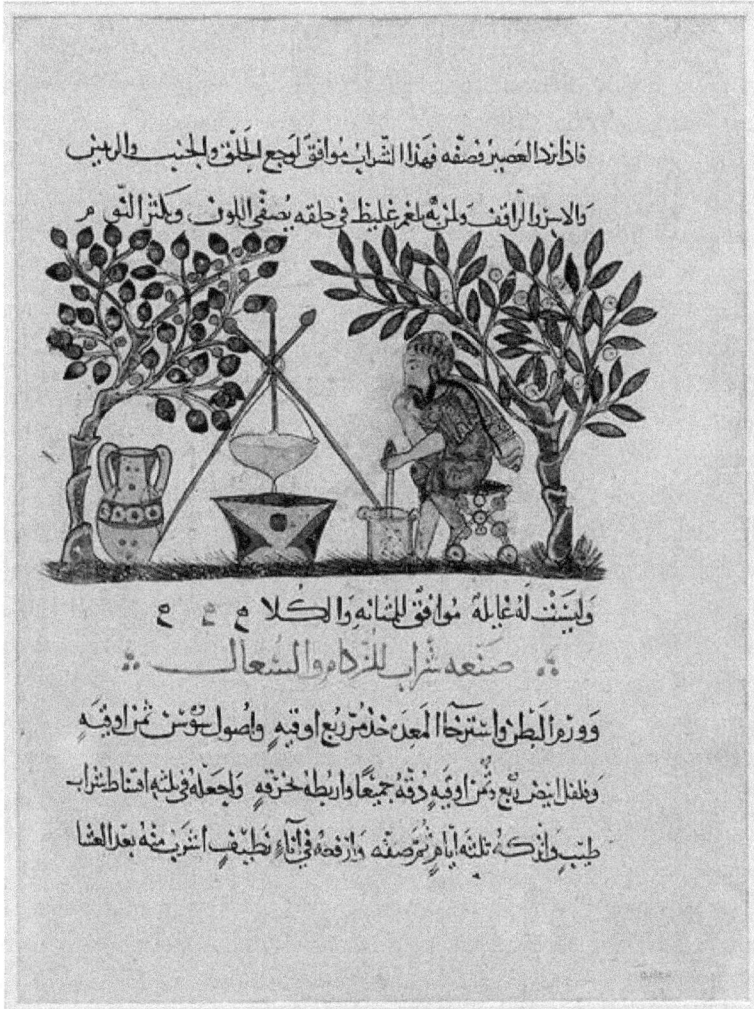

Physician Preparing an Elixir (40-90 AD). From Iraq or Northern Mesopotamia

CHAPTER 5

Ecodigital Art: Democratization, Globalization, and Interdisciplinarity

INTRODUCTION

Turning from digital archives to digital artworks, the aim of this chapter is to articulate the defining themes of digital art – namely, democratization, globalization, and interdisciplinarity – in relation to how artists work with technology to highlight the urgency of environmental sustainability and botanical conservation. While advances in engineering and technology are often touted as the solutions to the Anthropocene's ecological quandaries, few studies have concentrated on the interconnections between digital technology, artistic practices, and ecological sustainability. In its essence, "ecodigital art" (a term which I will put forward) crosses fluidly between art and technology with a strong underlying environmental ethics (see chapter 2 on reciprocity ethics). A potential transformative outcome of ecodigital art is the changing of public perceptions and behaviors concerning nature and humanity's fractured relationship to plant life. This chapter begins by defining and contextualizing digital art as a genre, in order to posit a growing engagement with environmental issues in digital artworks of recent years. The discussion will establish the groundwork for subsequent analysis of the use of living plants in contemporary artworks in chapter 6.

Some of the questions addressed here include: How do ecodigital artists negotiate issues of global culture as a progressively more powerful, homogenizing, and ecologically dangerous phenomenon? How are new media tools utilized by ecodigital artists in ways that call attention to and invite critiques of globalization and unsustainable environmental practices? To what extent are the shapes of ecodigital artifacts of different types determined by the technology they seek to critique? In particular, the works of new media artists Joseph DeLappe and Natalie Jeremijenko help to answer these questions by providing case studies of digital creatives adopting themes of environmental sustainability and engendering the artist as ecological activist. Although complete answers to these questions will not be possible in the scope of this discussion, I do proffer a framework for thinking critically about the intersection between art, technology, and environmentalism.

78

TOWARD AN UNDERSTANDING OF ECODIGITAL ART: THREE THEMES

The words "digital" and "sustainability" are ubiquitous components of today's vernacular in an increasingly globalized world. Yet, the two terms tend to be conceptualized as mutually exclusive with *digitality* generally associated with technology/culture and *sustainability* with nature/environment. The digital revolution of the 1990s introduced computer power to the public at an unparalleled rate (Lovejoy, Paul, and Vesna 2011, 2), but with minor concern for the ecological ramifications of technological advances. Within these developments, digital art emerged as a fluid set of artistic techniques, technologies, and concepts – often associated with the history of the computer but clearly lacking, in its early years, an environmental ethics.

Beryl Graham defines digital art simply as "art made with, and for, digital media including the internet, digital imaging, or computer-controlled installations" (Graham 2007, 93). However, digital art as it is known today has been subjected to a multitude of name changes, from computer art in the 1970s to multimedia art in the early 1990s to new media art more recently. The nomenclatural variety signifies that digital art and its naming are "characteristically in a state of flux" (Graham 2007, 106) – reflecting, in part, the mutability and evolution of the technologies and approaches used by artists. On the whole, the tendency to hybridize across media boundaries is characteristic of digital art productions. By shifting between media forms and employing a range of techniques, digital artworks resist the categorization of their works by genre or structure. Indeed, installation, film, video, animation, photography, internet art, software art, virtual reality projects, and musical compositions fall within its ambit (Paul 2003, 70).

A theme raised by digital technologies is the democratization of the arts. The ethos of democratization points to the belief that every person has the right to engage in the arts. A democratic view argues that all individuals should be able to explore their creativity and appreciate the artistic outputs of society. Thus, the democratization of art increases public access and involvement in artworks through a form of "regime change" (*e.g.* the development of new and potentially interactive platforms for art outside of "sanctioned" gallery or museum spaces). Art historians and sociologists of art measure democratization in terms of access, participation, interactivity, reciprocity, and decentralization. Paul DiMaggio and Michael Useem highlight the "increasing representation of nonelites among visitors to museums and performing arts events" as an indication of a growing democratic environment (DiMaggio and Useem 1989, 166). Improved public participation in the arts shifts the historical trend in which involvement was the domain of the middle and upper classes. Hence, democratization is a political, social, and, arguably, ecological process that aims to remove barriers to arts access and representation amongst rural communities, the working classes, the disadvantaged and those without university educations, hence fostering an "elite experience for everyone" (Zolberg 2003). Enhanced access to an artistic resource (*e.g.* a gallery, museum, exhibition, installation, object of art, creative process, medium, or material), along with the right to experiment and create, are the essential ideals of democratization.

The democratization of the arts through the emergence of digital technologies reflects the origins of critical theory. This body of thought interrogated the social value and roles of old and new media. In the late 1960s and early '70s, a call for public participation in democratic processes reflected the belief that social progress could be fostered through lateral, non-hierarchical, and two-way forms of communication. A strong critique of mainstream media channels (*i.e.* television, radio, and newspapers) entailed growing support for a new, democratic media that would involve a broader social and community base. The reconsideration of mass media was by no means specific to the 1960s. The Frankfurt School in the 1920s criticized mass media and advocated for social and political communications to operate in a more transparent and accessible manner (Lister et al. 2003, 43–44). The Frankfurt School was a group of scholars and critics based in Germany, including seminal philosophers and critics Theodor Adorno, Herbert Marcuse, Max Horkheimer, Walter Benjamin, and Jürgen Habermas. In the context of World War II and the social upheaval triggered by Fascism, the Frankfurt School developed "critical theory" through Marxist principles that advocated fundamental change of the world (activism), as well as critical analysis (scholarship).

Frankfurt scholars argued that the "culture industry" produced passive consumers rather than engaged, participating, and independent citizens (Lister et al. 2003, 386–387). With this basis in critical theory, throughout the twentieth century, the call to democratize media consisted of three aims: (a) the revival of community structures and the creation of a free public sphere of debate; (b) the liberation of communication from authoritarian control and threats of censorship; and (c) the experimentation with new forms of virtual community and the construction of identity as an active and ongoing process involving the input of the public (Lister et al. 2003, 70). Thus, in the context of critical theory, ecodigital art is an outcome of the democratization of art, striving to give voice to ecological issues and endangered environments, as we will see shortly through examples from the work of Joseph DeLappe and Natalie Jeremijenko.

Globalization is the second theme in which many ecodigital arts practices have evolved in relation to their associated media forms (*e.g.* the internet, teleconferencing, mobile telephony, social media). Globalization has been theorized as an intensification of the process of modernization at the root of ecological change and destruction worldwide (Christoff and Eckersley 2013, x). Known as a contentious theme within political science, economics, communications, and environmental theory, globalization can be defined as "a dissolving of national states and boundaries in terms of trade, corporate organization, customs and cultures, identities and beliefs" (Lister et al. 2003, 10). One outcome of globalization is an international economy produced by the activities of multi-national businesses, the emergence of global financial markets, and the increasing homogeneity of goods and services around the world (Lister et al. 2003, 194). Moreover, the emergence of "global culture" is attributed to globalization and, specifically, the world-wide influence of the American mass media (*e.g.* reality television shows, conservative online news programs, and Hollywood cinema motifs).

In his essay "Globalization and (Contemporary) Art," art historian T.J. Demos poses the following question: "How does artistic practice [...] define, negotiate, and challenge the cultural, economic and political forms of globalisation?" (Demos 2010, 211). In addition to "cultural, economic and political forms," one could also add "ecological" forms to Demos' enumeration, especially considering the negative impacts of globalization on the environment in terms of climate change and carbon emissions. If digital art is "made with, and for, digital media including the internet, digital imaging, or computer-controlled installations" (Graham 2007, 93), then what is the relationship between art and the processes of globalization (including the ecological) to which such media are inextricably connected?

New media has contributed (positively and negatively) to the processes of globalization by facilitating instantaneous digital communication that transcends regional or national boundaries. To conceptualize globalization as the outcome of digital media, however, invites the idea of "technological determinism" (Lister et al. 2003, 201). This position takes a variety of forms and strengths all of which broadly maintain that technology underpins the shape of human culture, society, values, and practices. The issue of determinism points to the dynamics between human users and technological devices created through innovation and scientific experimentation. Technological conditions affect everyday life and the construction of culture, to some extent (Lovejoy 2004, 311). On the one hand, "weak technological determinism" concedes that we ultimately have power over technology, even though technology always shapes our culture (Feist, Beauvais, and Shukla 2010, 5). On the other, a stronger version of determinism – or what is called "autonomous technological determinism" – asserts that there is the potential for humans to lose control of technology, that is, for digital devices to assume an almost autonomous form and to impact the world in unprecedented and, possibly, harmful ways (Feist, Beauvais, and Shukla 2010, 5).

Globalization underscores the issue of cultural production in the context of mass culture and mass media. Just as ecodigital artists probe the intrinsic democratic possibilities of new media for activist purposes, so too do they engage with the global interactive potential of internet technologies. However, polycentrism and decentralization offer countervailing perspectives. Both of these concepts point to the dispersion of power and the creation of new forms of community, including activist environmental ones. An alternative way to conceptualize globalization and its homogenizing effects is through polycentrism and decentralization. Polycentrism is a theory that the dynamics between the global and the local, the centre and the periphery, the north and the south, are as vital to consider as the broad-scale impacts of globalization and the traditional geographical centers of economic and cultural power (Western Europe, the United States and, more recently, parts of Asia). The concept foregrounds the dynamics between multiple, interrelated "sites," including physical locations, cultural positions, philosophical orientations, or aesthetic ideas (Scholte 2005). Ella Shohat and Robert Stam propose a "polycentric aesthetic" to encompass a diversity of "sites" and to call attention to artists and artworks existing at the thresholds between concepts, discourses, and identities (cited in Jones 2011, 169).

A polycentric perspective of a digital artwork examines the multiple positions that constitute a work – from the physical locations where collaborators are located to the theoretical, ethical, and aesthetic values and modes of the participants, including non-human, botanical ones (see chapter 6). The "local" dimensions of the work (*i.e.* the contribution of each geographical site to the artwork as a whole or the technological innovations forwarded by artist-engineer collectives with specific affiliations) figure into a polycentric interpretation of a digital artwork. Decentralization involves the decentering of established regimes (*e.g.* political, economic and, arguably, aesthetic and environmental) and the weakening of the control mechanisms of authority hubs. The networks spawned by new media have facilitated the process of decentering by democratizing access to information and modes of expression (Lister et al. 2003, 10). In terms of digital art, decentralization provides an illuminating perspective for analyzing works. How does an ecodigital artwork distribute authorship and creative authority across a widely-based network of anonymous participants? How does an ecodigital artwork decentralize the activities of artists, contributors, institutions, and actants (human and non-human) in a myriad of ways?

The third theme in this equation, alongside democratization and globalization, is interdisciplinarity. Is a ecodigital practitioner an environmentalist, artist, poet, scientist, engineer, conservator, botanist, or all of the above? This question points to the hybrid identities of artists and artworks in terms of interdisciplinarity. A term coined by social scientists in the mid-1920s, "interdisciplinarity" is the convergence of knowledge disciplines. The perspective reflects a broader momentum during the twentieth century to resolve the "two cultures" (*i.e.* art vs. science) dilemma in which the disciplines (*e.g.* the arts, humanities, biological sciences, engineering, etc.) were thought to limit the possibility of knowledge integration. Addressing concerns over specialization, interdisciplinarity entails the use of more than one discipline in an artistic practice. Its premise is that the disciplines collectively form the foundations of creativity and that, while individual disciplines maintain discrete identities within theory and practice, there is a degree of boundary-dissolution that is important to foster. There is "interdisciplinarity," in which disciplines collaborate to produce knowledge forms, and "transdisciplinarity," in which there is a deeper degree of integration and greater loss of disciplinary identity. Transdisciplinary artistic practice requires the methods and theories established in disciplines, and, conversely, disciplines need thought that is transdisciplinary in nature to go beyond the inherent limits of the discipline. Many ecodigital artworks are interdisciplinary or transdisciplinary in character insofar as they cross between art, science, engineering, and specific sub-disciplines (*e.g.* studio practice, biology, robotics, optics, etc.).

Interdisciplinarity is defined according to the degree of interpenetration between disciplines. Joe Moran (2010, 14) defines interdisciplinarity as "any form of dialogue or interaction between two or more disciplines." What is most essential to interdisciplinarity, according to Julie Klein (1990, 13), is a "dispersion of discourse" characterized by the placing of creative activities within a broader (*i.e.* not discipline-specific) framework. Allen Repko (2008, 6) describes the space between disciplines as "contested terrain." Environmental scholars stress the reality of engaged formal and informal interactions between disciplines (Soulé and Press 1998, 399). These theorists stress that interdisciplinary artists need to be conversant with the languages of other disciplines before, during, and after collaborative projects.

Furthermore, Roland Barthes (1977) asserts a more radical position that interdisciplinarity is greater than discipline-based knowledge converging to yield new forms. For Barthes, interdisciplinarity involves the transformation of disciplinary conventions:

> It is indeed as though the interdisciplinarity which is today held up as a prime value in research cannot be accomplished by the simple confrontation of specialist branches of knowledge. Interdisciplinarity is not the calm of an easy security; it begins effectively (as opposed to the mere expression of a pious wish) when the solidarity of the old disciplines breaks down. (155)

Expanding interdisciplinarity beyond its disciplinary basis, the neologism "transdisciplinarity" appeared in the 1970s in the works of psychologist Jean Piaget, sociologist Edgar Morin, and astrophysicist Erich Jantsch (Nicolescu 2002). In the nineteenth century, English polymath William Whewell's concept of "consilience" signified the interpenetration of knowledge "where disciplines are not juxtaposed additively but integrated into a new synthesis" (Walls 1995, 11). Building on Whewell's work, *Consilience: The Unity of Knowledge* by biologist Edward O. Wilson provides a contemporary interpretation of the interplay between the sciences, arts, and humanities. Wilson (1998, 8) defines consilience as "literally a 'jumping together' of knowledge by the linking of facts and fact-based theory across disciplines to create a common groundwork of explanation." In terms of ecodigital art, transdisciplinarity is applicable to a wide spectrum of research areas and creative practices. Ecodigital artworks are intrinsically transdisciplinary.

JOSEPH DELAPPE: *MAPPING THE SOLAR* (2013)

The three themes of democratization, globalization, and interdisciplinarity figure into the work of ecodigital artists who use new technologies and media forms to interrogate contemporary environmental issues, particularly species loss, climate change, non-renewable energy, and decreased water quality. The aim of the second half of this chapter is to apply the three core themes enunciated in the previous section to the analysis of ecodigital artworks from two prominent artists, Joseph DeLappe and Natalie Jeremijenko. As public, performative, or participatory works that integrate various technological forms into a system, the ecodigital productions of DeLappe and Jeremijenko make use of emerging technologies to call attention to environmental degradation and habitat abuses.

The three themes can be leveraged as a framework for analyzing other ecodigital artworks outside the scope of this chapter.

American artist Joseph DeLappe has been developing electronic and new media artworks for the last thirty years and is widely regarded as a pioneer of online gaming performance and electromechanical installation. DeLappe integrates online technologies into his art as democratic and accessible platforms for political and environmental protest. Although not an ecodigital work *per se*, his web-based project *Iraqimemorial.org*, launched on November 29, 2007, is a digital memorial to the civilians killed since the beginning of Operation Iraqi Freedom in 2002. The work can be described as:

> an online and physical exhibition of memorial proposals and projects dedicated to the many thousands of Iraqi civilians killed in the War in Iraq. Artists, designers, architects or other interested creative individuals or collaborators are invited to submit either proposed, imagined memorials or documentation of completed projects. (DeLappe 2011)

As a "culture jamming" nexus for international artists and designers committed to remembering civilian casualties of the Iraqi War, *Iraqimemorial.org* exploits the intrinsically networked qualities of the web, resulting in an open-ended, participatory, and interventionist work and practice.

A more explicitly environmental example from DeLappe's body of digital art is *Project 929: Mapping the Solar* (2013), a 460-mile, ten-day bicycle ride in which the artist dragged a large piece of chalk attached to the frame of his bike for the entirety of the trip. Depending on the viewer's distance, the uninterrupted chalk line could be perceived from the air as a faint trace on the ground. The environmental sketch resulting from the tracing of DeLappe's physical movements adumbrates a broad-scale, site-based ecopolitical artwork with a digital activist ethics. Indeed, the delineated area is large enough to accommodate the world's largest solar facility, which could meet the current energy needs of the entire population of the United States. DeLappe encircled US government lands where nuclear testing and military exercises were conducted, including the infamous Area 51 (the Nevada Test and Training Range and Groom Lake), Yucca Mountain (site of a major nuclear waste storage facility), and Nellis Air Force Base, all within reach of Las Vegas. The work's title refers to the 928 nuclear tests that occurred at the Nevada Test Site in the forty years between 1951 and 1992. Delappe's 929th "test" stands for the potential for sustainable technologies to satisfy the long-term energy requirements of nations but with few of the side-effects of nuclear or oil-based sources (DeLappe 2013b).

Project 929 incorporates the MMORPG software Blue Mars Lite in collaboration with Manifest.AR, an artists' collective focused on the application of augmented reality (AR) technologies to public art and activism. The post-performance AR version of the trip makes it possible for viewers to imagine the immense solar project envisioned by Delappe and The Union of Concerned Scientists. During the performance, the artist also made continuous use of GPS technology, as well as a bike-mounted digital camera for live streaming video and real-time documentation (DeLappe 2013a). As a site-specific work, *Project 929* implores viewers to consider the implications of globalization for energy consumption and, conversely, the merits of human-powered, bipedal travel. Like *Iraqimemorial.org*, the artist's use of commonplace digital technologies, such as GPS and streaming video, invites into its fold a broader base of users, only requiring access to the Internet, to engage with the performance. The interdisciplinary hallmark of the project is signified by DeLappe's dialogue with The Union of Concerned Scientists and his knowledge of the energy history and sustainable potential of the contested locations in which the performance is situated. The polycentric aesthetic of *Project 929* is both localized and globalized, consisting of multiple living and non-living agents. The work is distributed across time and space, but maintains a passionate local alliance with the disrupted desert areas of the American Southwest.

NATALIE JEREMIJENKO: *MUSSEL CHOIR* (2012)

While DeLappe's interventionist mode of artistic practice leaves only the faint and ephemeral imprint of chalk on pavement, other ecodigital artists produce, modify, or model whole biological-digital systems to convey messages of environmental sustainability. One such example of a public artwork with ecological overtones is *Mussel Choir* (2012) by Natalie Jeremijenko. An artist, engineer, and inventor, named one of the leading women innovators in technology (2011), Jeremijenko's previous work has addressed environmental and urban issues of water quality, food sustainability, and human health through public and performative modes. For example, she created the spoof Environmental Health Clinic at New York University where "impatients" were advised to adopt imaginative environmental practices for their benefit and for the good of the community. Moreover, her project *OneTrees* (1998–present) involved the planting of genetically clonal walnut trees in San Francisco neighborhoods. With the assistance of a plant geneticist and through tissue culture methods, Jeremijenko oversaw the cloning of one thousand walnut trees from the DNA of a single specimen. One thousand plantlets were installed at Yerba Buena Center for the Arts in sterile, sealed cups. Alongside the horticultural experimentation, a CD-ROM with software allowed users to create virtual trees on their computers through artificial life algorithms common to computer simulation games. The project reflects the tension between the material and the digital; carbon dioxide inputs are required to maintain both the living and virtual trees (see chapter 6 for further discussion of virtual plants). *OneTrees* is an example of a publicly engaged ecodigital art experiment that raises the same questions scientists and policy makers might be asking about technologies, while also allowing the public the opportunity to see the results first-hand, in this case, through the growth of clonal trees still growing in San Francisco neighborhoods (O'Rourke 2013, 235–239).

Mussel Choir similarly enables members of the public to monitor environmental conditions, but through a sonic symphony produced by mussels and their aquatic ecosystems. In 2012, a test version of the work premiered at the Venice Biennale, with permanent installations planned for the Hudson River of New York and the Melbourne Docklands. Supported by Carbon Arts and the Australian Network for Art and Technology (ANAT), the Melbourne installation uses mussels to amass information about and artistically represent the water quality of the Docklands habitat (Carbon Arts n.d.). The remarkable ability of mussels to filter water is gauged by hall effect sensors, normally used for electronic applications, such as speed gauging. The hall sensors measure the opening and closing of the mussel shells. The data is converted into sound patterns – which become the audible incantations of the mussels. The behaviors of the creatures correspond to the behaviors of the installation in which the activities of mussels in response to environmental conditions (*e.g.* seasons, weather, and pollution) are translated to sound. The project's audio mappings include sound pitch to water depth, timbre to pollution levels, and tempo to the shell's opening and closing. *Mussel Choir* highlights the fragility and significance of aquatic environments, making audible the elusive biological mechanisms of these irreplaceable aquatic creatures and their habitats. The public ecodigital work engenders awareness of ecological issues, connecting the biological to the digital with a clear ethical underpinning.

CONCLUSION

The three themes of democratization, globalization, and interdisciplinarity are crucial for understanding works of ecodigital art, which combine artistic practice, digital innovation, and environmental ethics. *Mussel Choir* makes acutely clear that human interdependence with other organisms is at the root of sustainability, while *Project 929* underlines the energy legacies of a place and the possibility for renewable sources in the future. Since the emergence of digital technologies in the 1990s, artists have experimented and, at times, pushed the boundaries of innovation (Hope and Ryan 2014). However, more recently digital artists have capitalized on the potential of new technologies to express environmental concern, to engage the public in thinking seriously about ecological issues, and, hopefully, to influence policy makers as another, powerful means for change. This triad of themes serves as a critical framework for understanding other works of ecodigital art, such as those presented in chapter 6, as they become increasingly more common online and in physical forms in relation to the growth of technologies.

CHAPTER 6

Plant-Art: The Virtual and the Vegetal in Contemporary Performance and Installation Art

INTRODUCTION: PLANT-ART IN CONTEXT

Whether as paintings, sketches, textiles, or craftwork, plants have been integral to Western art over the ages. Florilegia, herbals, botanical illustrations, pressings, and other renderings of whole plants and, very often, of their flowers comprise part of the tradition of plants in art. Closely related to these visual forms are tactile and sensory practices engaging plant materialities and involving, for example, natural dyes from roots, resins turned into adhesives, figurines sculpted from fine wood, and leaves incorporated into the texture of an artwork. A third dimension of the plants-as-art tradition regards the vegetal (*i.e.* tree, herb, orchid, flower, trunk) as intrinsically a living work of art – a complete botanical form not requiring visual rendering or material manipulation by humans to become an artwork. As such, a plant is *a priori* a paragon of natural beauty and an expression of harmony, symmetry, color, and other aesthetic qualities in itself. This latter aspect involves the appreciation of botanical nature on its own terms – in its raw state – without the intervention of an artist. These three elements of the tradition of plants in art – let us call them "visual plant art," "tactile plant art," and "plants-as-art" – have been transformed by the introduction of digital technologies into creative practices since the 1990s (chapter 5). Hence, there is presently a need to articulate a fourth element of the plants and art tradition – the subject of this chapter – which I will call "plant-art," with a conjoining hyphen signifying the inseparability of the two terms. Here, living plants – not necessarily the most aesthetically pleasing ones – are involved fundamentally as agents and active co-creators in a digitally-based work.

Involved in this manner in the becoming of a digital artwork, plants are neither the objects of aesthetics nor the subjects of a preservationist environmental ethos. Rather they are positioned integrally as contributors or "actants" in the world, bringing together the natural and the virtual within a single work (Hitchings 2003, Latour 1987, 2005, Tsing 2010). The plant-art work becomes a dynamic, shifting and organic locus or what Donna Haraway (2011, 12) terms a "terrapolis" – a "'niche space' for multispecies becoming-with" that is "open, worldly, indeterminate, and polytemporal" and rich in "materials, languages, histories, companion species" (see chapter 1 for further discussion of posthumanist theory). Despite the work of Haraway and other posthumanist scholars, the concept and practice of digital plant-art represent an interbreeding of technology, art, and plant biology that has received surprisingly limited scholarly attention (Nemitz 2000, Wilson 2002, 18-20). Meanwhile, there are numerous precedents for thinking about visual plant art (Blunt and Stearn 1950, Saunders 1995), tactile plant art, especially in Indigenous traditions (Clarke 2007), and plants-as art (Ryan 2012a, chapter 5). In order to articulate a fourth intersection of art and flora, the following chapter foregrounds the use of living plants as agents in digital art through a reading of several contemporary examples. The "co-becoming other" (Mules 2014, 22) of the virtual and the vegetal – of the artwork and the sensing and sentient plant – takes place when artists and audiences interact with flora during the creation of time-based, open-ended, or participatory works. More specifically, this chapter will consider five representative plant-based performances and installations, beginning with Laurent Mignonneau and Christa Sommerer's seminal visualization *Interactive Plant Growing* (1992) and concluding with Chiara Esposito's imaginative *The Dream of Flying* (2013).

Despite a range of possible conceptual approaches, including feminist science and technology studies (Haraway 2011, Hayles 2002), three posthumanist frameworks will be invoked in order to theorize aspects of human-plant-technology co-becoming, multispecies relationality, and bioartistic affect in plant-art works. These include digital theorist Roy Ascott's concept of "moistmedia" (Ascott 2000), Warwick Mules' notions of poiesis outlined in *With Nature* (2014), and the emerging field of inquiry, tentatively called "human-plant studies" (Ryan 2013, 72–89). The latter term describes the growing interdisciplinary dialogue between new, though contentious, scientific research into plant neurobiology (for example, Trewavas 2006) and emerging literary, artistic, cultural, philosophical, and legal approaches to plants (for example, Marder 2013a). Underlying these performances and installations is a perception of plants as interdependent in their environments – as autopoietic beings with nuanced sensory and decision-making faculties that enable them to participate in the "event" of an artwork (Mules 2014, 80). In addressing the epistemology of plant-art works – the means through which knowledge is produced through them – I will focus on three elements found in these frameworks: relationality, co-becoming, and affect.

DIGITALITY, MOISTMEDIA, AND HUMAN-PLANT STUDIES

The concept of plant-art developed here depends on a definition of digital art – "art made with, and for, digital media including the internet, digital imaging, or computer-controlled installations" (Graham 2007, 93) (see chapter 5). Broadly defined as art incorporating technological practices, digital art is "characteristically in a state of flux" (106). The flux results, in part, from the mutability and constant evolution of the technologies and approaches used by artists. The sheer diversity of names – often erroneously applied as synonyms for "digital art" – is further indicative of the flux and includes new media, multimedia, computer, software, hypermedia, emergent media, unstable media, electronic, and internet art. For example, internet art is based on the internet; browser art specifically makes use of internet browsers; and software art involves the use or creation of computer software. Other terms related to digital art – including behaviorist, interactive, and sound art – are more inclusive than "digital art" because they encompass both analogue and digital art practices, for example, ranging from site-based installation works to internet-controlled telerobotics. Still, other terms are period-specific and seem like anachronisms now. These include "net.art" which designates the digital art of practitioners working in the 1990s when internet technologies began to emerge in the public domain (Hope and Ryan 2014). Although technological in character, many digital artworks are also highly conceptual, interactive, open-ended, process-based, and, in the case of plant-art, organic, hybridic, and relational. The inclusion of real plants in artworks is an extension of the inherent flux of digital art as a genre and practice, and moreover indicates the variety of works signaled by the term (see chapter 5 for a more detailed discussion of digital art).

How might we characterize artworks that merge inert technological "material" (plastic components, memory cards, data flows) and actual flora (trees, shrubs, vines, herbs, common potted plants)? What does a plant become (*i.e.* how does it change or remain the same) when incorporated into an artwork as a biological interface between human participants and the digital apparatus of the work? The speculative writings of art theorist Roy Ascott offer a basis for conceptualizing the vegetal and virtual relationship (also see chapter 4). In particular, Ascott's "moistmedia" signifies the hybridization of the "dry" plastics and pixilations of the digital realm and the "wet" molecules and matter of the biological world within a new media artwork. In Ascott's terms, moistmedia (including plant-art works) lead to "edge-life" or new forms of identity, sociality, and aesthetics that combine digital and biological exigencies (Ascott 2000, 2). In short, edge-life entails the hybridization (rather than the polarization, if that were even possible in today's world) of the virtual and biological, the digital and material. Ascott further speculates that moistmedia "comprising bits, atoms, neurons and genes" leads to a "transformative art concerned with the construction of a fluid reality" and the "recognition of the intelligence that lies within every part of the living planet" (2000, 2). The virtual-vegetal dialectic is "technoetic" – the amalgamation of *techne* (in Heideggerian philosophy, a means of technical control, world shaping, and visual rendering) and *gnosis* (or spiritual knowledge about the world, culminating in enlightenment or insight). As indeterminate, fluid, and self-regulating systems, moistmedia prompt the rethinking of the very premises of intelligence, consciousness, and sentience:

> Moistmedia is transformative media; Moist systems are the agencies of change. The Moist environment, located at the convergence of the digital, biological and spiritual, is essentially a dynamic environment, involving artificial and human intelligence [and plant intelligence] in non-linear processes of emergence, construction and transformation. (Ascott 2000, 4)

"Moist" is a mediating category between the digital and biological, the immaterial and material, the inanimate and animate. In similar terms, Ascott (2000, 6) goes on to define "cyberbotany" as the intelligent application of plant "technologies" to the complexities of the biological-digital intersection and to the aesthetics of creative production in new media environments. Cyberbotany signifies the co-becoming of the digital and vegetal that typifies a work of plant-art or otherwise. Along comparable lines, biologist Stefano Mancuso uses the term "bio-inspiration" to refer to the application of plant intelligence to the design of computers and society, resulting in systems that are "networked, decentralized, modular, reiterated, redundant and green" (Pollan 2013, 104–105). The intelligence of the plants – circumscribed by their capacities for communication and behavior (chapter 11) – intersects with the *poiesis* (*i.e.* making or bringing forth) of the artwork. In plant-art works, the *autopoiesis* (self-making-becoming) of the plant underlies the poiesis (making-becoming) of the digital system.

In addition to moistmedia and bioinspiration, human-plant studies (HPS) presents a framework for investigating plant-art works and their contexts (Hall 2011, Marder 2013a, Ryan 2013). In previous work, I have outlined the premises and tenets of HPS as: (a) plants are always intelligent and volitional organisms; (b) plant intelligence is integral to socioecological networks and practices; (c) plant intelligence is a viable exemplar for human societies, cultures, and communities; (d) the roles of plants in society are best articulated through interdisciplinary research spanning art, literature, philosophy, Indigenous knowledges, and science; and (e) HPS complements but departs from the existing academic discourses of ethnobiology, ethnobotany, economic botany, and other cognate but different socially and culturally-based fields of plant research (Ryan 2013, 77). Although an interdisciplinary field, HPS is strongly underpinned by trends in "plant neurobiology" and research into plant cognition, acoustics, learning, and memory (Gagliano 2012, 2013, Gagliano et al. 2014, Trewavas 2006). HPS acknowledges that mainstream plant science tends to classify vegetal behaviors as tropisms – mechanical responses attributable to electrical or hormonal flows (*i.e.* auxin is often involved). These dominant science-based rationales for plant behavior do little to incite the rethinking of intention, consciousness, emotion, self-awareness, or cerebrality, root-bound or otherwise. Instead, they reinforce and privilege the Cartesian premise of the human subject with brain-bound intelligence. Indeed, part of the scientific resistance to plant neurobiology could be the moral implications of plant consciousness, which, if given cultural traction, could rouse the broad-scale reconsideration of vegetal pleasure, pain, and ethics (Hall 2011) (see chapter 11).

Nonetheless, with current science as its reference point, HPS explores the intelligent habitus of plants, including their "sessile life styles" and sense capacities in relation to human powers of smell, taste, sight, touch, and hearing (Chamovitz 2012) (chapter 12). These scholars – although located in different disciplines – hold that the reconsideration of plants in Western thought demands a "democratic" definition of intelligence (*i.e.* not restricted to mammals) as the potential to solve problems and adjust to an environment (Pollan 2013, 100). Moreover, following Hayles (2002, 4) and other posthumanist scholars of feminist science studies, a "distributed" concept of intelligence also recognizes that cognition is not circumscribed by the processes of the brain alone but rather is a "systemic activity, distributed throughout the environments in which humans move and work." A non-cerebrocentric mode of intelligence calls for a variety of human and nonhuman, animate and inanimate, actants, including artists, audiences, plants, and technological apparatus, all working in dynamic relation to their environment, including the gallery space and the interiority of their bodies.

HPS explores the philosophical implications of plant intelligence as "distributed" or "swarm' intelligence" – a category usually applied to social insects such as termites and bees to foreground the primacy of "the connections between the individual workers that form a network and changes in communication [that alter] the behavior of the whole colony" (Trewavas 2006, 7). Like the science with which it engages, HPS counters the "cerebrocentric" or animal-focused conceptualization that permeates our understanding of intelligence and situates the mind in the brain (Niemeier 2011, 48–49). Plant physiologist Anthony Trewavas – perhaps the most unapologetic proponent of intelligence in plants – uses the term "mindless mastery" (Trewavas 2002). Additionally, Michael Pollan characterizes a "leaderless network" (2013, 100) in terms that evoke Deleuze and Guattari's anti-hierarchical metaphor of the plant rhizome. Indeed, researchers across the humanities and sciences increasingly proffer the view that plant intelligence must be understood on its own terms – involving a considerable rethinking of the vegetal in Western thought that engages a sustained and critical examination of language (Vieira, Gagliano, and Ryan 2016). As a form of visual language, plant-art can be positioned within the critical human-plant studies context. Indeed, many plant-art productions call attention to different manifestations of botanical volition as plants decide – in a seemingly spontaneous and unchoreographed manner – to participate in (or opt out of) a work. As we will see later in this chapter, Natasha Myers (2013) problematizes the possibility of plant volition in art through the phenomenon of "lag." In sum, plant-art situates plants integrally in the design of the artwork, thereby circumventing anthropocentric problems of representation and aestheticization. Intelligent plants, therefore, become contributors to the creative arc of the digital work, especially process-oriented installations.

SENSORY ECOLOGIES: SEEING AND TOUCHING PLANTS

One of the salient features of plant-art, highlighted in this section, is sensory interaction between visitors and flora through the interplay of different sense modalities – both human and botanical. The following works of plant-art draw attention to the complex entanglements between plant and human sensoria, as well as plant and human existentiality. The elements of relationality, co-becoming, and affect that underpin plant-art works also influence the sensory ecologies of the installations and performances, such that the individuated sensoria of actants necessarily become collective, permeable co-sensoria impacting and configuring each other. A seminal plant-art work of this type is Laurent Mignonneau and Christa Sommerer's *Interactive Plant Growing* (1992) which used living plants as active interfaces between gallery visitors, the computer system, and the artwork (see Sommerer and Mignonneau 2011, for images of the work). The digital-vegetal installation, now held in the permanent collection of the ZKM Media Museum in Karlsruhe, exemplifies visual and haptic human-plant interactions. Participants could generate three-dimensional images of virtual flora when they touched or approached living plants. The visualization of botanical growth on the projection screen was moderated in real-time by human-plant encounters in the interior space of a gallery. The installation innovated the use of plants as a "natural and tangible interface" in a work of computer art, instead of commonplace "dry" technical interfaces, such as joysticks and keyboards (Sommerer and Mignonneau 2011, 206). Moreover, *Interactive Plant Growing* made possible a human-plant "dialogue" in which people participated in the creative process through botanical interfaces, resulting in an "open-ended artwork that is not predefined" (Sommerer and Mignonneau 2011, 206). As the artists themselves comment, "in 1992, using plants as interfaces to create awareness about the human-to-plant relationship by creating an artistic interpretation of this minute dialogue was very unusual" (Sommerer and Mignonneau 2011, 207).

The work involved five different living plants – a fern, ivy, cactus, moss, and small tree – placed in a semi-circle on five wooden columns directed toward a 12 by 9 foot video projection screen (Deussen and Lintermann 2005, 240). Measurements of the differences in electrical potentials – or voltage – between the bodies of plants and the bodies of visitors triggered the generation of images of floristic growth on the large screen. The voltage differential depended on the distance between a participant's body and the plants. Electrical charges registered when a participant's hand was between 0 and 70 centimeters away from one of the species. As visitors interacted with real plants in the gallery, the screen filled increasingly with a jungle-like visualization of ferns, ivy, grasses, mosses, and trees. The cactus served as an eraser plant, resetting the screen when touched. Each real plant corresponded to a virtual plant and to a discrete algorithm that generated a visual representation of its living correlate. In total, Sommerer and Mignonneau developed five algorithms reflecting the real morphological changes of the five plants over time in order to translate the people-plant electrical differential into tapestries of virtual growth. For each plant algorithm, there were several randomizing variables that determined the composition of the imagery – including stem length, width, curvature, branching angle, and color. As with other examples of digital or computer art, an algorithm developed by the artist-programmers controlled the translation process, in this instance, between the voltage differential, the variables, and the garden-like visualization.

This dialogic basis of *Interactive Plant Growing* – involving an ongoing feedback loop between plants, humans, and technology – can be theorized through the concept of "gesture." Gesture can be defined phenomenologically as "the openness of the body to the outside" during which embodied contact with the world is made (Mules 2006, 6). As "sensuous movements, leading to intimate encounters of touch, taste, smell, sound and looking closely" (Ryan 2012, 29), gesture disrupts the space of disinterested aesthetic appreciation intrinsic to many traditional works of visual plant art. Indeed, gesture is a core attribute of Mignonneau and Sommerer's installation and its open-ended and embodied aesthetic. It promotes a sense-engaged creative process, hybridizing the virtual and the vegetal in one "moist system." The simple acts of getting close to and feeling the leaves of plants are intimate sensory gestures that influence the composition of the garden-like imagery and infuse the digital system with the "wet" matter of the biological world. The installation invites visitors to participate – physically and creatively – in the ever-shifting arc of the visual artifact through their gestures toward plants, registered technically as voltage differentials between bodies in contact or approaching the moment of contact. Moreover, the participant's somatic influence over the information transmitted from the plants to the computer underscores issues of human-plant communication and biological feedback mechanisms. The moist installation depends completely on human-flora dynamics, which, in turn, depend wholly on simple human gestures toward plants. Interdependent with participants, the plants are "intelligent" collaborators (necessarily defined in non-zoocentric and distributed terms), rather than passive objects of representation (see chapter 11). As a whole, the installation takes the form of a self-regulating, open system capable of spontaneity and complexity. Thus, it exemplifies Ascott's assertion that "moist systems are the agencies of change." The change that results is both intrinsic (the constantly morphing garden visualization) and extrinsic to the artwork (the fostering, through art, of cultural perceptions of commonplace plants as creative agents).

A more recent installation by Mignonneau and Sommerer, *Eau de Jardin* (2004) – translated from French to "water garden" – was inspired by Monet's oil paintings of water lilies, created throughout his career but most prolifically toward the end of his life from 1920 to 1926. Based structurally on the panoramic layout of the Monet paintings at the Musée de l'Orangerie in Paris, the interactive installation was developed for the House of Shiseido in Tokyo. The work used a 36 by 9 foot, 3-sided projection screen to create an immersive triptych of a water garden, including a pool that reflected images of the virtual plants (Sommerer and Mignonneau 2011, 45). Ten transparent amphorae, hung from the gallery ceiling, contained aquatic plants, such as lilies, lotus, cypress, and bamboo, with their roots and soil visible to viewers through the glass of the vessels. The digital technology of the installation reflected their earlier work *Interactive Plant Growing* in which the voltage differentials between the visitors' bodies and the living plants were processed through a computer algorithm to generate plant images on the triptych screen. The computerized plants – populating the water garden – mirrored their living counterparts in the amphorae, while these plant images reflected on a virtual water surface. Unlike *Interactive Plant Growing*, *Eau de Jardin* involved a unique visual interlayering of virtual and reflected plant representations. However, in keeping with their previous work, the water garden shifted in composition according to visitor numbers and the intensity of human gestures toward the real aquatic plants. The translation of visitor dynamics in the gallery to renderings on the screen, again, highlights the conceptual and technical centrality of human-plant interaction in works of plant-art. As an expression of moist media, *Eau de Jardin* resulted in a "wet" environment (both biological and digital) crossing virtual and vegetal realities and merging artificial, human, and plant intelligences.

AFFECTIVE ECOLOGIES: SINGING, FLYING, AND PLAYING PLANTS

Interactive Plant Growing and *Eau de Jardin* connected visual and tactile interaction with real plants in a gallery to the algorithmic genesis of virtual plants on a projection screen. In contrast, the interactive installation *Akousmaflore* (2007), by artists Grégory Lasserre and Anaïs met den Ancxt working under the name Scenocosme, blends haptic, acoustic, and kinesthetic modalities (see Lasserre and Met den Ancxt 2014, to view images of the installation). Touching or simply approaching six potted plants hanging from the ceiling, a viewer triggers a symphony of acoustic loops, gradually increasing as visitor-plant interactions intensify. In response to different forms of gesture and movement, each hanging plant yields specific vocalizations, expressions, and textures. For example, the plants sing when touched or stroked. When hugged, some plants squeal. The ambient effect corresponds to the potted plants' excitation levels expressed by the voltage differential between them and human visitors. The artists describe Akousmaflore as an acoustic system hybridizing living plants and digital technology, and capitalizing on the ability of plants to function like "natural sensors" (Lasserre and Met den Ancxt 2014, para. 2). The integration of the natural sensing capacity of plants and the digital apparatus of the work echoes Ascott's concept of "cyberbotany" and Mancuso's "bioinspiration," discussed previously. Similar to the installations of Mignonneau and Sommerer, the algorithmic basis underlies a relationship between people, plants, and sound through the flow of data. Embodied experience in the gallery prompts the reconsideration of our perceptions of living things, particularly common houseplants which "have an ambiguous existence that swings between decorative object and living being" (Lasserre and Met den Ancxt 2014, para. 2). As with other plant-art works, *Akousmaflore* also exhibits a symbiotic co-becoming between digital technology and plant nature in the otherwise "denatured framework" of sensing instruments and computer equipment (Krajewski 2013, 138). This co-becoming is the open-endedness of the work and the basis of its epistemology.

Like *Akousmaflore*, *Singing Plants Reconstruct Memory* (2010), by Canadian new media artist Jo SiMalaya Alcampo, is an interactive, mixed-media plant-art installation in which sounds were produced in response to interactions with plants in a gallery (see Alcampo 2010 to view images of the performance). Alcampo, who grew up in the Philippines, meditates on her childhood traumas through the work's unique arrangement of potted plants, electronics, soundtracks, film, and video (Spooner 2010). The performance involved small banana plants connected to pressure-sensitive instruments. When the plants were touched, voltage passed into a grid controlling a soundtrack of traditional Filipino instruments and chants, as well as a projection of archival film footage from Alcampo's childhood. When watered, the plants "played" the music and footage continuously, articulating their higher levels of excitation. Wounded banana leaves were sutured by the artist, providing another somatic stimulus for the flow of sound and audiovisuals and connecting Alcampo's trauma to the pain (*possibly*) experienced by plants. Curator Rosie Spooner (2010) characterizes the performance as "memory work" – the unearthing of stories long concealed or repressed – in which plants are "silent witnesses" (or, in the case of this work, not-so silent witnesses) to trauma and the "keepers of cultural memory, muted histories and forgotten experiences."

For anthropologist Natasha Myers (2013), "lag" is pivotal to interpreting *Singing Plants* or, more precisely, to anticipating how the work *might* be received by an audience. A lag is a delay in response time – or sometimes a complete absence of response – between the moment of human interaction with the banana plants and the production of aural or visual effects in the artwork. As a gap in the call-and-response pattern, lag can be regarded as an expression of agency and subjectivity as the plant decides when to respond, participate, or perform. Alcampo's use of lag contrasts to the common scientific model of response which would regard lag as a distortion or anomaly in reaction to mechanical or chemical stimuli. Although she problematizes the premise, Myers (2013, 4) argues that "lags are read as recalcitrance, a plant's uncanny hesitation. It is in the space of this silence, a withholding or reluctance to respond to experimental demands, that plants seem to assert their agency." Myers (2013, 4) goes on to propose "moral and affective ecologies" or "the values, anticipations, feelings and desires that shape how and what plants get to express in their experimental configurations." When an experimental configuration involves the feedback loops of a moist system, feelings and desires take shape between species. Thus, visitors reported feeling satisfied after a banana plant responded: "when they talked to the plants they wished that there could be some feedback and that is what I find people are most moved by" (Alcampo 2010). The heightened awareness of plant volition – evident in *Akousmaflore* and *Singing Plants*– also underscores the non-zoocentric and distributed principles of intelligence that inform the field of human-plant studies.

Chiara Esposito's *The Dream of Flying* (2013), featured in the Ars Electronica Festival, responds to the common assumption that plant movements are slow, minute, and typically imperceptible responses to stimuli. A digital interface enables an air plant (*Tillandsia* spp.) and a common dandelion (*Taraxacum* spp.) to control their own movements on a small flying device (see Esposito 2013 to view images of the installation). Voltage differentials between the plants, visitors, and the gallery environment determine the movements of the plant's "body extension" or flying device. Like all works of plant-art featured in this chapter, the air plants and dandelions of Esposito's *Dream* are living sensors and natural interfaces between human participants and the open system of the digital installation. The salient quality of the work is its affective ecology of joy, playfulness, exuberance, and freedom that so markedly departs from the serious, sexualized, and opportunistic Darwinian plant constructed by botanical science for centuries. As such, plant intelligence becomes a viable exemplar for the digital artwork, as a moist system hybridizing the vegetal and the virtual.

CONCLUSION: EARTHING DIGITAL ART THROUGH VEGETAL AGENCY

The ethos of plant agency displayed in these works parallels the new scientific understandings of plant signaling and behavior research. These examples – in conjunction with the research they embody – counter the mechanistic understanding of plants as automatons, materials, or mere aesthetic features of landscapes. Indeed, writers and researchers from different disciplinary backgrounds have begun to critique the view of plants as "passive objects – the mute, immobile furniture of our world" (Pollan 2013, 94) that still permeates the traditional biological sciences as well as many Western cultural and humanities-based ideas of the vegetal. For Michael Marder (2013, 24), plant-thinking redresses the "systematic devaluation of vegetal life in Western thought" while also allowing for human encounter with plant difference. Marder (2013, 90) describes "vegetal existentiality" as "the time, freedom, and wisdom of plants [that] will come to define the positive dimensions of their ontology." Hence, thinking with plants is a mode of thinking that comes to reflect the particular habitus of the vegetal world, in contrast to reinscribing the zoocentric logos that dominates concepts of intelligence, sentience, and moral consideration. The critique of the devaluation of plant life also underlies human-plant studies in which plant-art can be contextualized.

However, in addition to prompting the rethinking of plants, the works featured in this chapter also signify a broader relationship between the digital and vegetal – between the immaterial and material, the technological and organic – that is poietic (see chapters 3 and 4). In relation to the poiesis of these works, digital plant-art offers an example of "earthing the world" in which art comes to bear a material relationship of becoming with the earth and with technology in an open-ended and indeterminate mode (Mules 2014). As such, the autopoiesis of flora – the becoming that is proper to plants in relation to their physiologies, motilities, sensorialities, temporalities, and ontologies – intersects with the poiesis of artworks, defined as the event of an artwork, including the processes whereby the work becomes (or is the constant state of becoming) through a feedback-driven process linked to its environment. In other words, the work of art and the habitus of the living plant are commingled and interdependent as part of the moist system of the digital artwork. Relationality emerges as plant-art actants influence and configure each other, shifting "the focus from entity [*i.e.,* the artist, audience member or individual plant] to relation [*i.e.,* the artwork as moist, open and intelligent system]" (Hayles 2002, 6).

Whether or not plant-art affirms or further calls into question contested notions of plant agency, intelligence, subjectivity, consciousness, and behavior is debatable and is not completely resolved by these examples. However, there is no doubt that further exchange between emerging plant science and works of plant-art will continue to stage vegetal existentiality and the "ontology of plants" (Marder 2013, 164), facilitating human encounters with, appreciation of, and ethical regard for plant difference in the midst of the relationality of moist systems. Indeed, posthumanist concepts of relationality and difference resist what we construct plants to be (*i.e.* passive and aesthetic objects), a determinate and zoocentric mode of thinking that has dominated Western epistemology for centuries (Ryan 2012, chapter 1). In highlighting and putting into practice the particular agencies of plants, the moist systems of plant-art also bear implications for vegetal ethics and politics (see chapter 2). Rather than the materials manipulated (dyes or fibers) and the objects fetishized (pleasing or symmetrical) of tactile and visual plant art, respectively, or the subjects dematerialized of plants-as-art, the air plants and dandelions of plant-art are actants, contributors, and participants embodied in the event of digital art. Hence, the biopolitical register of these works exists alongside other interfaces between vegetal life and technology, in which plant ontology is put to work or mirrored, including recent productions of "bio-inspiration" (Mazzolai and Mancuso 2013) and advances in "plantoid" robotics (Mazzolai 2010). To leverage a term from animal ethics, sentient plants are positioned as "subjects-of-a-life" (Regan 1993, 43) in works of plant-art, alongside other participants and actants (chapter 2). Nevertheless, one could argue that the plants are indeed conscripted, staged and, to some extext, manipulated for the creative intentions of the artists and audience. However, the foregrounding of the vegetal capacity for excitation and lag, consistency and mutability, vocalization and silence, calls attention to their subjectivities and resists their devaluation.

Moreover, the intimate involvement of living non-human beings in the digital infrastructure of plant-art works "earths" the overwhelmingly "dry" technological preoccupation of the genre since the 1990s (chapter 5). The plant-art installations and performances of this chapter – in contrast to techne – rematerialize art through the use of living things that always "refuse techne and affirm themselves as other" (Mules 2014, 197). And "by thinking with these things in their refusal, we open the technology [and I will add, the art] otherwise" (Mules 2014, 197). Nonetheless, the refusal of techne through the agencies and subjectivities of plants is conversely a symbiotic co-becoming – the moist system of the plant-art work the outcome of sustained feedback and dialogue between human beings, plants, and technology. The refusal is the open-endedness of the system of which the living plant is part. The renaturing of digital art through the use of living plants that can refuse – a renaturing involving the sensory and affective capacities of plants and people – preserves the "poietic openness" (Mules 2014, 11–12) of the works. It is this poietic openness that distinguishes plant-art from visual plant art, tactile plant art, and plants-as-art. Whereas visual, tactile, and plants-as-art forms exact degrees of representation or manipulation, plant-art produces a flux of meaning iteratively between the plant, artist, audience, and artwork in sensory contact. This flux is the basis of the co-becoming between Us and Other, between nature and technology, between the vegetal and digital, that is a salient mark of plant-art. The openness also casts a light on the potentialities of human relationships to the botanical world in the context of increasingly technological forms of art.

PART IV

Poetry & Vegetality

Page from Herbal

CHAPTER 7

"A Very Striking Parasite:" The Cultural History of the Western Australian Christmas Tree

INTRODUCTION: NUYTSIA FLORIBUNDA

I recently emailed a photo of the Western Australian Christmas tree, *Nuytsia floribunda*, to a Chinese friend in Nanjing. In uncharacteristically gushy fashion, she wrote back rapidly, "I like these yellow flowers. They are very beautiful like gold, like honey. I want to eat them!" Associating my flowers with hers, she then reminded me of *meihua*, the elegant plum blossom, the subject of much adulation in China.

Often we – in Australia – know more about the charismatic plants of the northern hemisphere than we do our own. Cherries, roses, tulips, oaks. But what lore of this brilliant golden tree of my part of the world could I tell my virtual friend? To venture an answer, I must begin with a premise: the Christmas tree is a perfect contradiction.

The cultural history of *Nuytsia* offers a glimpse into what makes Western Australia distinctive (chapters 3 and 4). Its legacy reflects the curious and sometimes dismissive, sometimes glowing, observations of naturalists, settlers, poets, artists and tourists – many of whom regarded (and still regard) the intriguing tree as a symbol of the isolated WA landscape and its anomalies, conundrums, and surprises (see Prologue).

The mellifluous name *Nuytsia floribunda* is of two inflections. The first bears the weight of Pieter Nuyts, the violent Dutch ambassador who mapped Australia's western coasts in 1626 on the Dutch East India Company's vessel *Gulden Zeepaard* (*Golden Sea-horse*). The crew recorded seeing *Nuytsia* near Walpole. However, I speculate that the first flashes of land spotted by an earlier Dutch mariner, Dirk Hartog, and those aboard the *Eendrachtsland* in early 1617, as they approached the coast of New Holland, were the young blossoms of the Christmas tree – burning yellow and orange iridescent against the earthen browns and ochres of the landscape.

The second inflection, *floribunda*, evokes the incandescent aura swallowing the tree for several weeks in the late spring and early summer in South-West Australia. This eco-region extends in a triangular shape from Shark Bay in the north-west to Israelite Bay in the south-east, close to the fabled ivory sand beaches of Esperance (see chapters 5 and 6 for more background). The South-West is recognized internationally as a biodiversity hotspot with almost half of its plant species, including the Christmas tree, occurring nowhere else. Such a rate of plant endemism is extraordinary, as previous chapters of *Posthuman Plants* have already outlined.

Nevertheless, few outside of Australia (and perhaps few within) would recognize a *Christmas tree* that is not an evergreen with a pleasingly tapering figure: rotund and earthward at the bottom, lean and heavenward at the top. But beauty oversteps the hard lines of biology, geography and culture. If beauty is a *language* (from *lingua*, the tongue), it can be tasted.

HISTORIES OF TASTE

My friend's synaesthesia aside, the WA Christmas tree is gorgeous enough to eat. Over one hundred years ago, Ethel Hassell became aware of this. In 1878, aged twenty-one years, she took up pastoral life with her husband Albert at Jarramungup, a sheep station north-east of Albany. On a late December, post-Christmas trip to socialize with neighbors fifty miles away, Hassell reported "in the distance on the plains a clump of the most beautiful tall trees covered with deep orange-coloured blossoms." At closer range, she and Albert observed Nyoongar people busily prying up the roots, or *mungah*, of *Nuytsia*, "tasting very like sugar-candy [...] sweet and more or less of a watery nature" (Hassell 1975, 26). Not only a quick (though laborious, by the standard of the minimart) carb kick, *mungah* were woody reservoirs during the long dry of the Western Australian summer. Sources of "beauty as well as bread" (to quote the early American nature writer John Muir), Christmas trees have been, for thousands of years, oases in these sandy plains hugging the Indian Ocean.

Sites of sustenance, sociality, sharing *and* sugar. Does this strike a familiar chord? Think about any cappuccino strip in any Australian city. Such conviviality continues around us, although in radically different and less healthy forms.

Indeed, to appreciate *Nuytsia*, and its sweet side, is to know the tree from the roots up – to avert one's gaze from the easy magnetism of its flowering canopy and to become subterranean, at least figuratively. One of the world's largest mistletoes, *Nuytsia* is half parasite, half plant. Its iron-grip rootlets, known to botanists as *haustoria*, feel discerningly in the earth for various hosts, while its leaves convert sunlight into carbohydrates as all good plants do. The tree is more accurately a hemiparasite – one that pilfers nutrients from other fibrous bodies (a liberal palate for a plant, from banksias to couch grass to utility lines) but also, as the early twentieth-century botanist D.A. Herbert put it, exerts "the power of independent existence after it has once become established" (quoted in The West Australian 1919, 6).

A hemiparasite, like *Nuytsia*, is a producer and consumer – an ecological prosumer. A perfectly adapted contradiction comfortably at home in the thicket of binaries we impose on the natural world, in order to make sense of it.

In 1919 Herbert became the first to confirm the uncanny parasitism of the Christmas tree – a subject of sustained debate amongst Australian botanists until then. Lugging his lantern slides and photographs into the WA Museum at a meeting of the Royal Society, Herbert demonstrated "the method of attack" – "two white fleshy arms start to grow round the attacked root in opposite directions from the point of contact" (quoted in The West Australian 1919, 6). When the two "arms" meet, they fuse, encircling the host root like a wire clamp. Herbert concluded that, through "simple osmosis" (rather than cellular synthesis between host and invader), the root's "tongue-like masses of tissue" furtively glean the food they are after through this process. But the haustoria never penetrate the host's deeper layers of wood, indicating that the nutritive materials siphoned off are supplemental, rather than primary, food sources. The evidence was well-received. Herbert's paper was applauded as "one of the most important that has ever been read before the society" (The West Australian 1919, 6).

I admire *Nuytsia* rootlets for their pertinaciousness. Although I have never been parasitized by haustoria (nor do I wish to be, but who knows what the future will bring), the whole scenario evokes the sensation of my father's jocose hands on my neck on a Saturday morning. As an electrician, his fingers were used to cabling, clipping, clamping, strapping, pinching, soldering, vice-gripping. His haustoria were steel-hard from years on the job. For the good fortune of my circulatory system, they never needled their way into my plasma. We could speculate that the host plant's sensation of being sucked on is somewhere between a vigorous massage and a blood transfusion.

In addition to banksias, utility lines are foci of the hemiparasite's oral fixation. Vincent Serventy in his book *Dryandra: The Story of an Australian Forest* (1970) provides this anecdote about the species' disconcerting culinary habits. For a space tracking station (presumably the Muchea Tracking Station, established in 1960, closed in 1964), underground cables two centimeters in diameter were buried thirty centimeters in the sandy soil and insulated with a polymer coating to thwart fungi, termites, acids, and other agents of decay. "All went well until six months later. Somewhere the cables had short-circuited. The engineer raised the cable and found encircling it rings of white flesh" (Serventy

1970, 106). The carnivorous clamping mechanism was the same that Herbert had described about fifty years earlier with lantern slides. To the dismay of NASA's Project Mercury, the haustoria nibbled the plastic sheath, taste-testing the cables inside, but probably experienced a case of hemiparasitic indigestion. Serventy quips that *Nuytsia* would have suffered "some disgust, one imagines, as there would be little nourishment in those messages from outer space" (Serventy 1970, 106).

If we are to become a "sustainable society" with vibrant regions for all species, we need to laugh *with* (and sometimes *at*) ourselves and our flora, fauna, and fungi. The natural world is full of sex, comedy, and irony, especially when faux parasitism is involved. Serventy's minor bout of floristic humour reminds me of this. Biodiversity conservation is often too serious an endeavor – too macabre and dispiriting (see chapters 3 and 4).

But to label the haustoria *non-discriminating*, as many botanists do, is wrong-headed and verging on the ecologically impolite. Although seemingly misplaced at times, the root system is discerning, opportunistic, and ardent. Considering that it evolved in the mid-Eocene – roughly fifty million years ago – sampling of the inorganic is to be expected. According to botanist Stephen Hopper, *Nuytsia* came about following a period of high sea levels when the coast of the South-West consisted of islands and peninsulas (Hopper 2010). Adapted to the infertile and weathered soils of the eco-region, the endemic flora, including the seasonally flamboyant Christmas tree, has thrived through isolation. Buried utility lines, in comparison, are newbies to the scene, the first telegraph line in Western Australia being laid from Perth to Fremantle in 1869. In another fifty million, *Nuytsia* could figure out how to digest space-aged polymers. If intelligence entails, amongst other things, the capacity to evolve and adapt, then we have a genius in our midst (see chapter 11 for further discussion of plant intelligence).

In the Nyoongar language, the tree is *mudja*, a marker of birok – one of the six seasons in the traditional calendar of the South-West. Nuytsia is intimately connected to the afterlife and, according to the early twentieth-century ethnographer Daisy Bates (1992, 153), has been considered "sacred for its spiritual memories." As Noongar elder Noel Nannup explained to me, the Christmas tree's ecology and spiritual significance are interwoven: "A spirit sits on the tree until it flowers. Then the spirit moves on to the spirit world in conjunction with easterly winds and fire, which take the spirit out over the sea."

Colonial Perspectives on the Fire Tree

Earlier this year, I interviewed centenarian botanist David Goodall (born 1914), who immigrated to Australia from England in 1948. In his university office (he still goes to campus each day), I asked, what were your first impressions of Western Australia? Without blinking an eye: "the Christmas tree, a very striking parasite."

Perhaps you have noticed already. *Nuytsia* is not your run-of-the-mill Aussie tree like, say, the gum (no offense, eucalypts). Since the founding of the Swan River Colony in 1829, its combination of striking beauty and physiological peculiarity have flummoxed observers and has figured into its naming. Settlers came up with colloquialisms – many descriptive, some humorous, a few poetic (see Prologue).

Fire tree expresses the visual radiance of the species, framing the abundant blossom as a harbinger of burning. The golden profusion corresponds to the onset of the South-West bushfire season. John Lindley in *A Sketch of the Vegetation of the Swan River Colony* (1839–40), the first substantial published account of the Western Australian flora, comments that "such is the abundance of the orange-coloured blossoms, that the Colonists at King George's Sound compare it to a tree on fire; hence it has gained the name of 'Fire tree'" (Lindley 1840, xxxix). Lindley never visited the Colony (unlike the contemporaneous naturalist James Mangles), but, instead, relied on the previous accounts of botanists and cartographers, as well as word-of-mouth from settlers.

Other scriveners recorded *cabbage tree* as an olfactory, tactile, and visual moniker for *Nuytsia*. One would expect that the pungent, slightly fetid smell of the tree's cut wood stung the nostrils of pastoralists as they cleared the vegetation. Just a few years ago my own sniffing glands were overtaken by a whiff of broken Christmas tree as I tramped around a suburban Perth bushland reserve. Mild sinusitis and a vague feeling of euphoria ensued. Strangely pleasurable. In 1846, the settler George Fletcher Moore (1846, 80) referenced the term "cabbage tree" along with its Nyoongar and scientific names in the following: "Mut-yal, s.-Nuytsia floribunda; colonially, cabbage-tree. The only loranthus or parasite that grows by itself. Another anomaly in this land of contradictions. It bears a splendid orange flower." *Loranthus* refers to the showy mistletoe family, the Loranthaceae.

But the exact categorization of the "anomaly" into a plant family evaded morphologists during the nineteenth century. Philippe Édouard Léon van Tieghem, a colleague of Louis Pasteur, created the family Nuytsiaceae in 1896.

But ten years on, Ludwig Diels re-classified the Christmas Tree as a genuine Loranthaceae. Our tree still stands here today.

The algologist William Harvey in an 1854 letter to *Hooker's Journal of Botany and Kew Garden Miscellany* mentions *Nuytsia*. Concerning its parasitism, he thought "it highly probable that there is underground attachment to the roots of other plants" (Harvey 1854, 219). Like other botanical writers, Harvey liberally garnished scientific assessment with aesthetic impressions and folk knowledge, specifically the tree's unusual likening to one of the world's healthiest foods. But he strikes me as a bit cool and restrained – like a kelp, and unlike other more effusive *Nuytsia* commentators. "It is a very deformed-looking tree at best, but gay enough when in blossom; its leaves, too, are of a very beautiful tender green. They call it the Cabbage-tree" (Harvey 1854, 219). *Cabbage* could either refer to the tenderness of its leaves or the ease with which the axes of settlers penetrated its pithy wood. The metaphor could also imply a modicum of stinkiness. These names persisted throughout the nineteenth century. For now, let's take all three senses of the figuration.

The most effusive colonial-era observer of our tree would have to be Marianne North. The peripatetic botanical artist produced a very fine painting of *Nuytsia*, now held at Royal Botanic Gardens, Kew. In "Study of the West Australian Flame-tree or Fire-tree" (circa 1880), a gracefully fluted and pleasantly mottled trunk trifurcates about halfway between the earth and a crown of deep green foliage and golden tufts. A juvenile *Nuytsia*, aflame in flower, peers almost curiously from behind the spot in the stately tree where its trunk divides into three arms. To the right, a fiery bough cantilevers near a large gray boulder set amongst some seriously cute bonsai-like balgas (*Xanthorrhoea preissii*, also known as *grass trees*). North placed the elegantly composed *Nuytsia* – albeit curiously elm-like – at the crest of a rise, aside a foot track that descends to the vast floor of the purplish plains in the distance. This is most likely somewhere along the Darling Scarp, perhaps during North's overland trip to Perth shortly after her arrival at Albany.

I have been on the lookout for North's Christmas tree for the last six years, since I came to Western Australia from the northern hemisphere. But I have yet to find such a symmetrical specimen in the wild or anywhere else. Instead, *Nuytsia* comes in all shapes and sizes, from spare-looking, wind-sculpted bushes near Lucky Bay in the region's south-east corner to dazzlingly unkempt hemiparasites near Perth Airport on the Swan River coastal plain. I agree with Harvey somewhat, most are "very deformed-looking" trees, but their anomalous growth habits make them alluring, captivating, surprising. Their bark is dark and rough, and, on older trees, fractures into a raised pattern of small crevasses and rectangular chunks.

North recounts her memory of *Nuytsia* in *Recollections of a Happy Life* (1892), an impressive two volume memoir detailing her travels from Canada, the United States and Brazil to India, South Africa and Western Australia. Surprisingly I found an original copy of this work in the stacks of a local university library, not sequestered in a special collections area. It bears the elegant and thoughtful touches of a book from this era, including on its cover an imprint in golden ink of *Nepenthes northiana*, or the tropical Miss North's pitcher plant endemic to Borneo. In Volume II, she deploys the colloquialism, *the mistletoe tree*. "I shall never forget one plain we came to, entirely surrounded by the nuytsia or mistletoe trees, in a full blaze of bloom. It looked like a bush-fire without smoke. The trees are, many of them, as big as average oaks in our hedgerows at home, and the stems are mere pith, not wood" (North 2011, 153). This curious pithiness later led to the cutting down of Christmas trees for recreational purposes and sparked the outrage of conservationists, the wood of the species "a resilient, though durable, target for steel darts" (Sunday Times 1940, 7)

In the 1920s Emily Pelloe noted some of the folk beliefs surrounding *Nuytsia*. To pick its flowers before Christmas day is unlucky. To use it as a wedding decoration is to bring misfortune to the bride. A few of these beliefs continue today. Like the anti-dart contingent, who opposed the harvesting of *Nuytsia* wood for the manufacturing of dartboards, Pelloe (1926, 5) also pleaded for the conservation of the world's "most gorgeous botanical spectacle."

THE CHRISTMAS TREE: CONTEMPORARY CHALLENGES

I watched a lone *Nuytsia*, transplanted from the northern suburbs of Perth where its habitat was cleared for housing, languish for several years on the campus of my university, until one day it disappeared, removed surgically and silently by the grounds staff. It is known that uprooted Christmas trees rarely survive – their underground universes too sensitive, their ancient logos not readily translatable to another locale. Meanwhile, a stone's throw away, at the local golf course, they irrupted all through early summer like miniature suns, "like a fire in the woods," as James Drummond (quoted in Hopper 2010, 341) remarked in the 1800s. It had been a sodden spring season. In remnant patches of vegetation in suburban Perth, hemiparastic miracles go on as they have since the mid-Eocene, despite bushland clearing, vandalism, and disease.

A *Nuytsia* specimen bloomed at the Sydney Botanic Gardens in 1842, but died in 1883. Maybe North was aware of this. She finishes off her glowing recollection with a fitting disclaimer that "they have never succeeded in cultivating those trees in captivity" (North 2011, 153). Pelloe (1925, 5) concurs regarding "the hopelessness of its artificial propagation." This saga played out during Georgiana Molloy's short life. On the behest of Mangles in 1839, Molloy took up the task of collecting seeds – an arduous preoccupation that would last until her death in 1843. She wrote apologetically to the Captain, conceding that "I have been four times out in quest of Nuytsia and send you the small, small harvest. They are very difficult to obtain, if not there the very day they ripen" (quoted in Hasluck 1955, 241)

Molloy's labours were much later memorialized by Alan Alexander in his beautiful poem "Nuytsia Floribunda:"

> The parasite Floribunda for my drowned son.
> How delicate they are, these stars at random. (Alexander 1979)

Stars at random.Georgiana, her lost son and her elusive seeds bring home to me that this striking parasite is a heritage that can too easily be wiped out by unchecked "development" in the South-West and the global impacts of climate disturbance (chapters 3 and 4). The irony nowadays – to my mind – is that the fiery blossom continues to symbolize what is unique about Western Australia while the uniqueness becomes increasingly threatened. State identity through endemic flora is not a new story. It sits uncomfortably alongside the destruction of plant life.

In 2010 at the request of the Premier's Department, Jan Pittman's illustration of *Nuytsia* was featured as one of Colin Barnett's sanctioned Christmas cards. Yet hundreds of *Nuytsia* were expunged in the summer of 2013 to make way for the expansion of Perth Airport. I feel a pang of loss tinged with revulsion as I pass by where they once were, along the Great Eastern Highway, in all likelihood the first living specks of color noted by eastern states and international visitors flying into Perth for the holidays. The "not threatened" conservation status of the tree cannot tell this story.

Recently the Christmas tree has been selected by Earthwatch Institute as an indicator species (Ashbolt, Quaife, and Ryan-Charles 2012). Their ClimateWatch initiative involves large-scale data collection by citizens in an effort to demonstrate the ecological impacts of climate change. Although the results are inconclusive, one aspect has been made clearer: *Nuytsia* is a highly resilient species that, for nearly fifty million years, has evolved with a fire-prone South-West landscape. And other species rely on it, just as it depends on others. For instance, *Nuytsia* is the only suitable nesting tree for the yellow-rumped thornbill (*Acanithiza chrysorrhoa*) several years after a bushfire.

CONTEMPORARY CHALLENGES

The "very striking parasite" is a state of mind, a mode of consciousness in Western Australia. It's an aureate tuft in peripheral vision. It's a sting in the nose, a sweet watery sap. It's beauty and bread. North got *Nuytsia* right – it occupies the center, but also wafts in from the margin. It's a perfect contradiction in an imperfect world.

CHAPTER 8

Australian Ecopoetics Past, Present, Future: What do the Plants Say?

INTRODUCTION

And I came to a bloke all alone like a kurrajong tree.
And I said to him: "Mate – I don't need to know your name –
Let me camp in your shade, let me sleep, till the sun goes down."

–Randolph Stow, "The Land's Meaning" (2009, 248, ll. 30–32)

The previous chapter touched upon the role of poetry in the appreciation of plant life. Chapter 8 continues this discussion through the theoretical framework of ecopoetics. Like the country's arid interior, contemporary Australian ecopoetics is vast and robust. The expressions of Australian ecopoetry are as varied as the antipodean landscape itself, underscoring the intricate connections between language and ecology in this part of the world. The Mediterranean climate of Western Australia's South-West corner, the Red Centre of Uluru, the tropical rainforests of Queensland, the temperate Tasmanian old-growth forests and the alpine reaches of the Victorian High Country signify this: rather than a contiguous desert or a *terra nullius* (as some readers both inside and outside of Australia may *still* believe), the Australian environment is a mosaic of biota, climates, topographies, and regions.

And, moreover, each region brandishes a distinctive ecopoetic history in which writers have grappled with radically unfamiliar (to *their* eyes) plants, animals, fungi, and landforms through poetic exertion. Comparably wide-ranging, Australian modes are miniaturist, formalist, experimentalist, dystopianist, and activist as well as Modernist, Taoist, Romanticist, and Transcendentalist. Contemporary styles range from intimate meditations on nature as a dynamic phenomenon to the broadly-sweeping gestalts of the landscape poetry tradition. If I had to say it in one word, *plurality* defines Australian environmental poetry today. There are poets with ecopolitical motivations, some with spiritual sensibilities for nature and others for whom local landscapes germinate, as a matter of course, in their work as writers (and not necessarily as "nature writers" or "ecopoets") – as witnesses to what the country is and what it can *flower* into.

In what follows, I will be guided by my own green biases and botanical proclivities. To foot-slog a swathe through the intricate ground of Australian ecopoetics, I will don chlorophyll-streaked glasses – will listen closely to what the plants *say* (see also chapter 11).

LISTENING TO PLANTS

It's springtime in Western Australia, and I'm trying to hear the plants. If ever they should talk, I trust it will be this afternoon. Eager to bend my ear, I've followed a procession of cars towhat seems like a clandestine spot in the bushland fringes of the Perth suburb, Joondalup. I'm looking for a new way of looking at (and listening to) Australian flora – different to the what's-your-Latin-name-and-chemical-makeup-please approach. The sun breaches the gumleaf carousel overhead. Years ago, in America, I went to a lecture on communicating with plants. We each listened closely to a corn kernel, but I could only hear the pulse in my neck and my own unchoreographed borborygmi. Years later, I would find this same dark yellow and nearly fossilized kernel in the strangest spots: the ashtray of my car, the bottom of a suitcase, the dark hollow of a shoe in the morning. Now I wander in what I expect will be an effusive bush, listening for a vegetal sign that summons. I'm willing to wait (at least until the end of the New Age workshop) – but I feel untutored.What willthe call sound like? How will I know? How much will my imagination and interpretation need to play a role?

Seeking a familiar face (as if desperately awkward at a party), I squat lotus-style near a Mangles (or red and green) kangaroo paw (*Anigozanthos manglesii*) – an easy conversant I hope, the floral emblem of Western Australia since 1960. Endemic to the South-West of the state, this flower is one of the first to irrupt out of the sodden spring earth – a harbinger of spring, and simply avery attractive flower, its architecture perfectly adapted to pollination by probing bird beaks. As I go poodling around, skirting other spinally-erect meditators, its chlorophyll-glistens, its young-as-lettuce leaflets are coming up – blood-red splotches suffusing a photosynthetic vibrancy. This gregarious sprite of the bush duff.

The poet and conchologist William Hart-Smith (1911–1990), who lived in Western Australia in the late 1960s, memorialized the iconic red and green flower in this short, imaginative verse:

> A Kangaroo Paw
> by the roadside

with scarlet trousers
is thumbing a lift
with a vivid green thumb. (Hart-Smith 1979)

Hart-Smith captures the perceptual signature of the plant: the scarlet-to-green juxtaposition. His playful personification of the kangaroo paw "thumbing a lift" is as compact as a haiku or a roadside plaque. The jocular *femme* flower spoke to him, although gesturally, not verbally. Interspecies communication, indeed, seems to have happened between Hart-Smith (or at least between his car) and the blossom. Presumably these "conversations" still take place (especially for some New Age excursionists). Dangerous anthropomorphism? *Pathetic* pathetic fallacy? Can ecocritical over-thinking taint the pure delight of the poetic well? Is hearing plants speak to us a matter of our own belief or disbelief? And how might poetry become a medium for hearing – an interface between the "speaking" us and the "unspoken" other?

Attention to the things of nature blends our utterances, images, and metaphors with the language(s) of things themselves. The twenty-first entry in Australian poet Robert Gray's work "Illusions" puts faith in things (immediacy) over representations (meditation). He asserts the fallacy of believing "that it is not actual things we perceive, in their minute differences, uniqueness, and subtlety, and with such spontaneity and surprise, but representations of them only. (That this distinction can mean something)" (Gray 1993, 66, ll. 29–32). I too trust in the blurring of the distinction. Hart-Smith did. He seemed ticklishly surprised by the kangaroo paw. Yet others throughout Australian history have been less surprised, that is, more befuddled by the differences and more invested in the distinction.

For early observers, antipodean species inverted the conventions of nature recorded elsewhere on the globe (see Prologue). Trees that shed bark in summer. Large marsupials that hop. Monotremes, like the platypus, that lay eggs. Seasons, the diametric opposite of those in the northern hemisphere. In 1793, the British botanist James Edward Smith opined that naturalists "can scarcely meet with any certain fixed points from whence to draw [...] analogies." He proclaimed the plant life of the New World "total strangers, with other configurations, other economy, and other qualities" (Smith 1793, 9–10). With similar sentiment, in 1825 the judge and poet Barron Field lamented the absence of the four seasons in Australia as well as the tried and true natural pillars of poetic inspiration, decreeing peevishly "I can therefore hold no fellowship with Australian foliage, but will cleave to the British oak through all the bareness of winter" (Field 1825, 424).

Poets and other observers would continue to wrestle with the strangeness of the colonial landscape, attempting to breed familiarity through linguistic flourishes. Examples of catachresis are common: Swan River mahogany (not even close to a mahogany, but a eucalypt species endemic to WA), sheoak (no cousin to Field's British oak, but rather a casuarina), hoop pine (not a pine relative at all), and grass tree (lumped initially into the lily family, but, as it turns out, neither a grass nor a lily, and rarely growing as large as a tree). Concomitant to these intrusions of language were introductions of biota: rabbits, foxes, camels, bees, cane toads, creeping lantana, cape daisies, and the list grows on and on.

What does this all have to do with contemporary Australian ecopoetics? I suggest that the postcolonial context today brings about two impetuses in poetry. The first is the engendering of language that resists the residues of colonial-era predeterminations and attends now to the things of the Australian environment as they are, rather than how we think they should be (or indeed have been). This forges a new groundwork of ecopoetics – "points from whence to draw analogies" – material anchors for language, through the fusion of commitment to ecoregionalism, study of Indigenous environmental knowledges, competence with natural science, interest in the human senses, and appreciation of the dynamism of flora and fauna – beings and places – "down-under." The second is the engendering of an ecological ethics through poetry that brings attention to the compromised situation of nonhuman species and the habitats we share with them in an era of exponentially hastening environmental change (Steffan et al. 2009).

Known for her involvement in the campaign to protect the Great Barrier Reef and Fraser Island in Queensland, Judith Wright listened reverentially to plants. Her orchid writings make use of their scientific names – often historically-telling, poetic in their own right, and mellifluous to the lips. Consider "Sun-Orchid:"

> Sun-orchid, Thelymitra,
> what a blue of blues you've chosen
> to remind this sullen season
> that still the sky is there. (Wright 1994, 414, ll. 1-4)

The taxonomic name *Thelymitra* for the sun orchid genus was devised by naturalists on James Cook's second Pacific voyage to Australia (1772–1775). The term stems from *thely* for "woman" and *mitra* for "mitre hat," in reference to the intricate configuration of its reproductive structures. The poem is a direct address to an orchid – the use of the second person singular – with an emphasis, in the ensuing stanza, on the orchid's multi-sensorial milieu, and specifically the intimate linkage between human memory and olfaction "when a eucalyptine vapour / dreams up in windless air." Religious inflection comes through in the metaphors "wrapped up in your Mary-blue," "blessed from your creation," and "a gold like revelation." Another notable example of Wright's botanical poetry is "Phaius Orchid:"

> Out of the brackish sand
> see the phaius orchid build
> her intricate moonlight tower
> that rusts away in flower. (Wright 1994, 88–89, ll. 1–4)

Wright's rendering recognizes agency in things, in which the orchid actively "builds" and "weaves." The sightless orchid, or "blind being," is also a metabolic creature, transforming "sand's poverty, water's sour, / the white and black of the hour" into an image of natural beauty witnessed by the curious poet – an image neither aestheticized as an object of art nor rationalized as an object of science. The poet and the plant make contact between categories in a liminal space of unknowing knowingness. Wright attunes to the poietic temporality of the botanical world, in which the retreat of "the gift as soon as made" is part of an ancient cycle weaving time, life, and death. In short, she is thinking with plants.

According to Michael Marder, plant-thinking comes in these forms: (a) the "thinking" that is specific to plants or "thinking without the head" (what plant scientist Anthony Trewavas cleverly calls "mindless mastery"); (b) human thinking about plants, evident, for example, in poetry such as Wright's; (c) the moderation of human thinking by plant-thinking, through attunement to such phenomena as plant time, also a theme in Wright's botanical work; and (d) the ongoing and long-term interconnections between plant-based human thinking and plant beingness (Marder 2013b). The implications of plant-thinking for ecopoetics, in my view, are far-reaching (see chapter 6).

> There appear rarely
> the improvised strange contraptions
> of the trees –
> they are like crazy antennae. (Gray 1993, 26, ll. 21–24)

A litmus test for the evolution of Australian ecopoetics is the plant (as well as the fungus – and possibly the rock and the soil). The "sessile lifestyles" of flora demand an attitude of being with even the craziest, the most laconic and miniscule nonhumans – those regarded by many through the ages as "passive objects, the mute, immobile furniture of our world" (Pollan 2013, 94). Extending sensorially to plants through poetry is a gently radical act. It can reconfigure Romanticist notions of "landscape aesthetics" and Enlightenment notions of "natural science" dominating the Western ecopoetic canon at least since Wordsworth. Unless you are walking through the sublime karri forests of the South-West of WA, the ancient beech rainforests of Queensland, or underneath the hulking Huon pines of Tasmania, the "strange contraptions" of Australian flora are, for the most part, unassuming at a distance, especially flowers and leaves. Long bouts of heat and drought, solar and wind exposure, nutrient-deprived soils, and grazing by herbivores have caused the majority of plant species to miniaturize.

A postcolonial, posthumanist ecopoetics of plants is about paying attention – and learning how to listen, a process whereby the botanical becomes a lens for the literary, rather than vice versa. In a mimetic sense, the shapes of plants – rhizomatic and arboreal, lateral and horizontal, ecological and modular, cyclical and poietic – impart form and flow to poetry, and underpin a kind of ongoing reflexivity between landscape and language in a place. This is especially evident in the Aboriginal Australian traditions of "singing" country – briefly discussed later in this chapter – which involve human language to encourage fecundity in nature during certain seasonal phases.

Thinking about my innocent bush séance in Joondalup, why should the outwardly benign notion of hearing nature's voice be considered fallacious or dangerous? The example of casuarina (sheoak or swamp oak, mentioned earlier) is revealing. For nineteenth-century poet Charles Harpur, the voice of casuarina was mute yet sorrowful, lowly yet preternatural:

> Up in its dusk boughs out tressing
> Like the hair of a giant's head,
> Mournful things beyond our guessing
> Day and night are utterèd.
>
> Even when the waveless air
> May only stir the lightest leaf,
> A lowly Voice keeps mourning there
> Wordless oracles of grief. (Harpur 1973, 53, ll. 5–12)

In Harpur's poem, the oak's "voice" is not of internal provenance to the plant, but is rather generated mechanically by the external movement of the wind through its "boughs, dark intermassing." Here, Harpur recapitulates Marcus Clarke's infamous and long-lingering assessment from 1876 of the Australian bush as "weird melancholy" (Clarke 1993). In 1872, Western Australian settler Janet Millett did a little better than Harpur and Clarke, observing on her way to York "a few weird shea-oaks destitute of leaves, between whose fine countless twigs, doing duty for foliage, the air sighs in passing with the sound as of a distant railway train" (Millett 1980). All three colonial-era writers concur on the "weirdness" of the plant, but somehow Millett's sigh resonates in a less prejudiced and more nuanced way than Harpur's moan, at least to my ears. To her credit, Millett's observations of the sheoak's "foliage" are, still, ecologically accurate. But think of the plant subject: the lowly casuarina! Addressed historically by another's name; accused consistently of creepiness and black magic; attributed wrongly as a origin of settler depression in the bush.

Whose voice(s) are we hearing anyway? Nature's? That of our forebears? Our own? I suggest that part of the challenge of contemporary ecopoetics is to hear nature anew, to learn to listen and write in a different way – one which decouples the very concept of "voice" from human-based definitions, one which appreciates the diverse sensory expressions of "language" and agency in the natural world without succumbing to age-old biases (chapters 11 and 12). Therein lies the art of environmental writing today.

With these ideals in mind, my poem "Sheoak Reverie" aims to vindicate the casuarina. Still, there's so much work to be done on this Australian plant and others:

> lore hunts us down the same,
> Nantosuelta lurking on the plain
> feminine oak or the settlers' bane;
>
> tiny teeth are your verdure
> neither as leaves nor as needles
> but as cladodes, unlike the pine. (Ryan 2012b, 17)

The interpenetration of words and things (rather than the imposition of the former on the latter) is one of the concerns of contemporary ecopoetics. Indeed, many Aboriginal Australian Dreaming stories present a view of land as a living fabric comprising language and all that exists. Aboriginal country is often sung into existence – each of its cyclical changes inspirited by poetry and song. For example, the Narangga people of South Australia *sing* the ripening of wild peaches or quandongs: "Wild peaches hanging in the trees, the sun will burn you (to the colour of fire), we will gather you (for food)" (Clarke 2007, 24). Another way to think about this interrelationship between language and landscape is through the concept of *poiesis*. We know that the root of the word "poetry" is *poiesis* for "making" or "producing." For environmental philosopher Warwick Mules, *poiesis* "identifies the being of things in their becoming other: in their creative, shaped and connected possibilities" (Mules 2014, 22) (see also chapter 6 on plant-art). Mules goes on to describe a relation of "co-becoming other" between a creative work and a living being (Mules 2014, 22). This echoes Gray's questioning of the hard-and-fast distinction between things and their representations. Other scholars use the term "sensory poiesis" as the potential to "enact, rather than merely represent, the immediate, embodied experience of nonhuman nature" (Knickerbocker 2012, 17).

Of all Australian poets, Les Murray's work most effectively triangulates poiesis, language, and plant life. For instance, "Flowering Eucalypt in Autumn" (2007) ripples with movement and life, at a time of year when a tree becomes an ecosystem unto itself:

> That slim creek out of the sky
> the dried-blood western gum tree
> is all stir in its high reaches:
>
> its strung haze-blue foliage is dancing
> points down in breezy mobs, swapping
> paceand place in an all-over sway. (Murray 2007, 65, ll. 1–6)

Murray accomplishes a rare sense of ecological dynamism through his linguistic twists and turns. The poem's internal zing mirrors that of the flowering eucalypt: a co-becoming other. The active participle ("dancing," "swapping," "crisping") riffs off of poietic phrasing ("all-over sway," "night-creaking," and "fig-squirting"), sustaining the scene's procession – far from fixed and far from purely visual. The result is a language of sensuousness and specificity that matches that of the plant, the season, the moment. The emergence of the gum flower is:

> as a spray in its own turned vase,
> a taut starburst, honeyed model
> of the tree's fragrance crisping in your head.
> (Murray 2007, 66, ll. 19–21)

Taste, touch, smell – the intimate, the bodily, the autocentric. The limbic, the primordial, the mnemonic – the most exacting and most difficult to devise language for. The senses (and the struggle to bring them to language) mark contemporary ecopoetics (chapter 12). While visually evocative of Australian plants, Murray's poetry is haptic, olfactory, gustatory. WA poet Andrew Lansdown also handles the non-visual exceptionally well. In "A Few Weeks Later I Returned To Find" (1979), it is a certain grasstree that consumes him:

> it was a powerful, honey-thick
> nectar. The odour was a heavy
> sweetness. I wiped the pollen from my nose.
>   (Lansdown 1979, 8, ll. 21–23)

Like Murray's flowering eucalypt, Lansdown's grasstree is not just part of a habitat: it is a habitat, continuously shifting before his senses.

> my life, I imagined, must be a hymn
> to the optic nerve.
> Other senses, you have proven,
> will have all they deserve.
>   (Gray 1993, 20, ll. 73–76)

Besides an openness to the minutiae of experience, contemporary Australian ecopoetics evolves through the regional sensibilities of poets, exemplified by John Kinsella's work in the Wheatbelt of WA. As a localized call-and-response, hearing the voice of plants entails dwelling physically in their places, dwelling sensuously alongside them. In "Habitat" from his collection *Armour* (2011) we find a shift within the poem from speculative abstraction ("grave clarity," "a meal of distance," "authorial silence") to the shock of sensory immediacy in the bush:

> But that's okay,
> I say so myself, slipping over in mud
> and cutting my arms, face, on scrub.
>   (Kinsella 2011, 9, ll. 11–13)

The crystallization is an ecoregionalist poetics of place ("a miniscule patch"), plants ("trees"), the elements ("water," "sun"), and the audible ("sing out"):

> Trees
> too wide where water runs, gone, quick as the sun
> Here, a miniscule patch you sing out from.
>   (Kinsella 2011, 53, ll. 16–18)

Whereas the title "Habitat" is a generic (like the cognate terms "ecosystem," "landscape," or "environment"), Kinsella's "Resurrection Plants at Nookaminnie Rock" is botanically particular and place-specific. Nookaminnie: a rock outcrop near Quairading in the Wheatbelt of WA. Resurrection plant: a pincushion lily (*Borya* spp.) capable of enduring prolonged dehydration. In the thick of the poem's contemplation of death, life, birth, and rebirth emerges an ethics of plants based on regional ecological knowledge and the recognition of the limits of embodied encounter:

> Step carefully around these
> wreaths hooked into granite sheen, holdalls
> for a soil-less ecology, a carpet you know
> would say so much more if your boots
> were off and skin touched life brought
> back, restored, gifted, bristling with death.
> (Kinsella 2011, 53, ll. 14–19)

Record-hot summers. Record-low rainfall. Record-few wetlands. Record-sprawling housing estates. Much of the history of Australian environmental poetry has concerned the process of "coming-to-terms" – the formation of cultural identities in relation to the character of the biota here. But wild plants and other species are becoming intellectual specters rather than tactile presences and forms, as a consequence of climate disturbance, megapolization, plant disease, species displacement, and natural attrition. Nowadays bush melancholy takes a shape Harpur and Clarke wouldn't recognize. The philosopher Glenn Albrecht terms this contemporary form of environmental despair "solastalgia" or the feeling of homesickness while one is still at home – as the place where one lives rapidly changes and deteriorates, and as the species one has affection for, including kangaroo paws, orchids, eucalypts, and resurrection plants, become less common (Albrecht 2005) (chapter 1). And I don't think we can write quickly enough to keep pace with the juggernaut of natural systems decline. If we follow the idea of ecopoiesis as "co-becoming other," then the loss of plants is the loss of poetry, in Australia and elsewhere. All I can do is to keep listening and singing out from my own "miniscule patch," learning to hear what the plants say (chapter 11).

CHAPTER 9

Dharmic Ecologies Down Under: An Ecocritical Perspective on Buddhist Symbolism in Australian Poetry

INTRODUCTION: DHARMIC ECOLOGIES

How has Buddhist symbolism been used by Australian poets to represent the antipodean landscape, including its plant life? Indeed, a small but robust segment of contemporary Australian poetry alludes to Buddhist motifs, as will be shown through examples from the writings of Randolph Stow (1935–2010), Robert Gray (b. 1945), and John Mateer (b. 1971). Some of the symbols are ancient and canonical, some are the results of immersion in the Australian context, and others are hybrids borne of the poets' imaginations. On the whole, their works reflect an obvious and somewhat sustained interest in Eastern symbolisms but in explicitly Australian settings: the bush, the ocean, the desert, the forest. More specifically, the portion of their poetry that has been inspired and shaped by Buddhist doctrines invites an encounter between Western and Eastern poetic forms, philosophical precepts, and physical locations. Focusing on the latter (that is, the environmental and place-based dimensions of Australian Buddhist poetic symbolism), I will consider the convergence between antipodean landscapes and Buddhist symbolism in their poetry in this chapter. The connections between Buddhist ecopoetics in Australia and the concept of the posthuman plant will be addressed.

Furthermore, adopting an ecocritical framework, I will foreground the role of Buddhist symbols in environmental consciousness in Australia, as well as in expressing, in poetic form, the particular features, qualities, and experiences of Australian landscapes, with attention to its plant life. This constitutes a phenomenological approach to interpreting Buddhism symbolism in Australian poetry. In particular I will apply Peter Jaeger's concept of a "Buddhist ecopoetics" through an analysis of references to eco-dharmic concepts in select Australian poems. In the works highlighted for this discussion, consciousness of and multisensory attentiveness to antipodean environments – including oceans, forests, deserts, waterfalls, animals, and plants – underpin a nexus of ecologies, places, moments, ideas, and symbols. In other words, for these poets, the local environment becomes the material terrain for poetry, place, and spirit or, in the

122

words of literary critic Kevin Hart, the "field of dharma." Buddhist symbols are thereby translated to new forms involving Australian nature as their reference points.

BUDDHISM IN NORTH AMERICAN AND AUSTRALIAN POETRY

Even a cursory survey of contemporary poetry in English will reveal the influence of Buddhist ideas and practices. This is evident in verse – either tacitly (through close reading and interpretation) or ostensibly (readily apparent from the most superficial glance at the work) – and is time and again invoked in a considerable body of poetry. Most conspicuously in North America since the mid-twentieth century and in particular for the Beat Generation and post-Beat American writers, Buddhism has had a distinct bearing on the form and subject matter of contemporary poetry. In North America, Buddhist-inflected poetry has a lively tradition emerging from distinctly American permutations of Buddhist thinking and doing. Indeed, the influence of Buddhism is most apparent in the work of the Beats, the post-World War II writers who emerged during the 1950s and included influential literary figures such as Diane di Prima (b. 1934), Allen Ginsberg (1926–1997), Jack Kerouac (1922–1969), Gary Snyder (b. 1930), Philip Whalen (1923–2002), Lew Welch (1926-1971), and others. The wax and wane of the Beats followed their popularity in the 1950s and their eventually being subsumed within the countercultural movements of the 1960s. After this era, in May 1987 near San Francisco, a gathering called "The Poetics of Emptiness" at Green Gulch Farms Zen Center brought poets together for a weekend to contemplate the relationship between writing, meditation, and daily life (Schelling 2005, xiii). The poets represented different generations – from those born in the 1920s to younger writers born in the 1950s. The Green Gulch gathering was a defining moment in American Buddhist poetry and invigorated more concerted scholarship into Buddhism and poetry.

Shortly after Green Gulch, seminal studies and anthologies began to appear, notably *Beneath a Single Moon: Buddhism in Contemporary American Poetry* (Johnson 1991). As the first significant collection describing of the impacts of Buddhism on North American poetry, the collection includes over 250 poems and 30 essays from 45 poets, including Beat fixtures Ginsberg and Snyder, as well as younger poets like Jane Hirshfield (b. 1953) who were influenced by Buddhism but not as Beat Generation writers *per se*. Parallel to the anthologizing of Buddhist American poetry, scholarly research into the transmission of Buddhist ideas and practices to North America began with studies such as *How the Swans Came To the Lake: A Narrative History of Buddhism in America* (Fields 1992). In

1998, an anthology titled *What Book!? Buddha Poems from Beat to Hiphop* featured poetry again from Kerouac, Snyder, and Whalen, as well as Maxine Hong Kingston (b. 1940) and other writers not necessarily featured in previous collections (Gach 1998). Still, a few years on, *America Zen: A Gathering of Poets* (McNiece and Smith 2004) and *The Wisdom Anthology of North American Buddhist Poetry* (Schelling 2005) were published, the former including the work of Di Prima, Hirshfield, Snyder, Whalen, and other familiar names. More recently, *The Emergence of Buddhist American Literature* (Whalen-Bridge and Storhoff 2009) examines the influence of Buddhism on North American literature and the role of literary works in the reception and retention of Buddhist ideas in the continent. During the last ten years, studies of particular schools of Buddhism and individual poets have been completed, such as the monographs *Allen Ginsberg's Buddhist Poetics* (Trigilio 2007) and *Han Shan, Chan Buddhism and Gary Snyder's Ecopoetic Way* (Tan 2009).

In comparison, unlike its North American counterpart, the writing of Buddhist poetry and its scholarship in Australia is largely a more recent phenomenon with fewer poets, poetic works, studies, and anthologies. Of early note is Australian poet Harold Stewart (1916–1955), who followed Shin Buddhism, lived periodically in Japan, and became a scholar of Oriental culture and thought. Stewart is arguably the first Buddhist Australian poet, but many of his works take place in Asian, rather than Australian, settings. The anthology *Windchimes: Asia in Australian Poetry* is the nearest equivalent to the collections of North American Buddhist poetry enumerated above, but focuses on the more general relationship between Asian culture and Australian poetry (Rowe and Smith 2006). The anthology includes the work of about fifty Australian poets, such as Judith Beveridge, Robert Gray, and Jan Owen. To date, scholarly studies of the influence of Buddhism on Australian poetry have been either journal articles or book chapters focused on the aesthetic and stylistic facets of Buddhist symbols, for example, as mentioned briefly in Paul Kane's chapter "East-West Turnings"(2010). Also of importance is Kevin Hart's article "Fields of Dharma" (2013) which considers the Buddhist dimensions of Robert Gray's long poem "Dharma Vehicle." Hart's trenchant and well-informed analysis draws parallels between "Dharma Vehicle" (1978) and T.S. Eliot's "Four Quartets," both poems devoted to the principle of dharma, in the Hindu and Buddhist senses, respectively. Of further note in Hart's article is his brief ecocritical analysis of the interconnections between the coastal setting of Gray's poem and the poet's invocation of dharma. He states, on the sense of temporal passage – the transience of things – characterizing the poem, "there is a consolation: the immediate presence of things, especially nature. And, for poets and their readers, there is a further consolation, one for which the poem is also a vehicle: the sensuous description of nature" (Gray 2013, 282). I will return to these points – the field of dharma and the sensuous perception of nature – when looking more closely at Gray's poem later in this chapter. However, I now turn to the nature of the Buddhist symbol itself in relation to dharmic principles and the potential for it to translate to new settings, such as the antipodean landscape and its plant life. I will also introduce the framework of ecocriticism for interpreting Buddhism symbolism in relation to the natural world of Australia.

BUDDHIST SYMBOLS AND DHARMIC ECOLOGIES IN AUSTRALIA

A religious symbol is a representation of the beliefs of a religion as a whole, or of a precept within a religion. While often a single object, a symbol can constitute a system of interrelated signs. Indeed, Buddhism on the whole proliferates in symbols of different forms and modes – from golden icons to fragrant incense to the sounds of drums and bells. Moreover, while we tend to think only in terms of visual icons, symbols can be verbal or aural (chanting mantras), tactile (writing on prayer fabric), olfactory (incense), or gustatory (ceremonial food or drink). John Powers (2007, 241) observes that, in Tibetan Buddhism, a wide range of symbols is presented in order to accommodate (and indeed encourage) the different understandings and interpretations of individual practitioners. Some of the predominant and even well-known Buddhist symbols include the Eight Spoked Wheel, the Bodhi Tree, Buddha's Footprints, an Empty Throne, a Begging Bowl, and a Lion, all often used to represent Buddha. In particular, the Eight-Spoked Wheel or Dharmachakra (*dharma* as law or truth, and *chakra* as wheel) is a symbol for the Wheel of Truth turned by the Buddha. The Dharmachakra consists of eight spokes symbolizing the Eight-fold Noble Path. Additionally, the Bodhi Tree is another highly visible Buddhist symbol, representing the tree under which Buddha achieved enlightenment, with the tree leaf as a potent symbol of religious faith and devotion (Powers 2007, 258). For Buddhist practitioners, there are three primary categories of symbolisms and symbolic actions: *mudras* (hand gestures), *mantras* (words), and *mandalas* (icons) (Blau and Blau 2003, 6).

It is, however, critical to bear in mind that symbols take various sensory forms and, hence, that there is a phenomenological aspect to understanding symbolic meanings. This consideration becomes important as we look at the work of Stow, Gray, and Mateer and their use of the non-visual senses in relation to the Buddhist symbols with which they engage and, in fact, devise, construct, and transform through their poetry. For in Buddhist Australian poetry, as Hart comments, "we pass from otherworldly religious thought to direct perception of nature" from "being oriented to another world to embracing this world in all its sensuous particularity" (2013, 278). The nature of this passage (from the other world to this world, or from the sublime to the material, from abstract thought to embodied presence) is an ecocritical concern (or, a concern that can be illuminated through ecocritical analysis). One catalyst for landscape symbolism in Buddhist Australian poetry is direct sensory contact with nature. It has a relationship to dharma and, also, a register within ecological consciousness, awareness of one's place, and concern for the wellbeing of nonhuman life there. Despite the potency of canonical symbols such as Dharmachakra, I am interested in new symbolic permutations, as Buddhist ideas and practices circulate in Australian settings and gestate in the material terrain. As such, typically Australian habitats and biota – such as gum trees and paperbarks – take on dharmic meaning and become vehicles in themselves for Buddhist understanding, learning, and teaching.

The Sanskrit term *dharma* (or *dhamma*) is described by different writers and philosophers in a myriad of ways: as the state of nature perceived through the senses, as a phenomenon of nature and its qualities, as the laws of nature, or as the teachings of Buddha on the laws of nature and their implications for the human condition. The late Venerable Buddhadāsa Bhikkhu (1906–1993) outlines a four-fold definition of the Sanskrit *dharma*. He critiques the commonplace definition of the term as referring to the Scriptures, what he calls "the dhamma in the bookcase," or the verbal explanation of the Scriptures by a teacher. He asserts that this everyday understanding is that "of a deluded person who has not yet seen the dhamma" (Buddhadasa 1989, 128). Therefore, dharma as such does not simply refer to manuscripts or teachings, but to something that is at once profoundly material and stable yet highly immaterial and transient. This is nature as *dhammajāti*, or the intricacy of all things perceived by one's senses while remaining, simultaneously, outside the faculties of our perception. For Buddhadasa, dharma encompasses "nature itself; the laws of nature; a person's duty to act in accordance with the laws of nature; and the benefits to be derived from acting in accordance with the laws of nature" (1989, 128).

Buddhadāsa's concept of dharma offers a stimulus for environmental consciousness. Especially if Buddha entreats humanity to behave in accordance with natural laws and if there are discernible benefits in doing so, then dharma is inherently connected to ecology and environmental thinking. Sensing the intrinsic connection between ecology and dharma, the Beats and other Western poets and philosophers praised (and still praise) the natural virtues of Buddhism for environmental balance, or what we would today call "ecological sustainability" (see chapters 5 and 6). The eco-dharmic aspects of Buddhadāsa's teachings are more clearly evident in his political concept of "dhammic socialism." He explains that nature (or dhamma) involves a dynamic equilibrium between humanity, animals, plants, and other forms of life, such that there are resources available to all living beings. Nonhuman organisms tend to produce and consume what they need – without stockpiling, hoarding, and other forms of blatant excess – achieving a natural state of balance. The birds, insects, and trees enact, without the intentionality and ideology that come to characterize human thought, a form of ecological socialism. The scenario that Buddhadāsa outlines is an embodied, corporeal, and sensory one, reminiscent of the nature writing of nineteenth-century Americans Henry David Thoreau (chapter 12) or John Muir, for example. He implores us to:

> Look at birds: they consume as much as their stomachs can hold. They cannot take in more than that. They have no granaries for hoarding. Look at the ants and insects: that is all they can do. Look at the trees: they can take in only as much as their trunks will allow. Thus, this system, in which no being was able to trespass upon another's rights or hoard what belonged to others, is natural and automatic, and that is how it has been a society and continued to be one, until trees became abundant, animals became abundant, and human beings became abundant in the world. The freedom to hoard was controlled by nature in the form of natural socialism. (quoted in Puntarigvivat 2003, 191)

Hence, for Buddhadāsa the laws of nature (or *dhamma*) concern interspecies balance, communities of beings, equitable distribution of materials, the avoidance of excess for the benefit of all forms of life, and the value of abundance over scarcity. Dharma is, therefore, not only a spiritual concept but an ecological one; it is this manifold of meaning that Australian poets convey in their landscape writing.

  In order to develop an ecocritical perspective on Buddhist symbolism in Australian poetry, I will first distinguish between three interrelated terms: ecocriticism, ecopoetry, and ecopoetics. To begin with, American scholar Cheryl Glotfelty formulated the first definition of ecocriticism nearly twenty years ago as "the study of the relationship between literature and the physical environment [which takes] an earth-centred approach to literary studies" (quoted in Garrard 2004, 3). However, Glotfelty's definition strikes us now as somewhat outdated on a few counts. Most notably, recent modes of ecocriticism not only limit their analysis to literature (poetry and prose) but to works of cinema, performance, music, gaming, and popular culture, not excluding online materials (*e.g.* blogs) and other non-literary forms. Moreover, recent trends in material ecocriticism, zoocriticism, and botanical criticism have begun to critically reconsider the role of the living environment (including soils), animals, and plants, respectively, rather than the "physical environment" only. As ecocriticism finds it origins in the work of Marxist scholar and key early figure in cultural studies Raymond Williams, accordingly the field enacts a "political mode of analysis" (Garrard 2004, 3) in keeping with other forms of cultural criticism (that is, other "isms"). In sum, there are two primary focal points of ecocritical analysis: the representation of nature in texts and other media; and the degree to which cultural materials (including "texts" defined broadly) reflect notions of ecological justice, ethics, sustainability, and the voices of Indigenous peoples.

Ecopoetry is an object of ecocriticism; in fact, a poem or a poet can be identified as having an environmental imperative following some form of ecocritical analysis of a writer's work, or of the writer him- or herself. Scott Bryson outlines three features of ecopoetry, beginning with "an ecocentric perspective that recognizes the interdependent nature of the world" (Bryson 2002, 5–6) or, in Timothy Morton's terms, reflects ecology as "thinking how all beings are interconnected, in as deep a way as possible" (2010, 255) (chapter 1). Secondly, as Bryson (2002, 5–6) argues, ecopoetry expresses "an imperative toward humility in relationships with both human and nonhuman nature." And thirdly, ecopoetry reveals an abiding suspicion of "hyperrationality and its resultant overreliance on technology." In Bryson's view, ecopoetry is, at the same time, ecological, literary, and political. Whereas ecopoetry refers to ecological poetry, the term *ecopoetics* comprises the making (the techniques and the overriding philosophies informing the poetry) and study (the scholarship and interpretation) of poetry about landscape, wilderness, animals, plants, and environmental injustices (Skinner 2015). It is possible, then, to speak of a Buddhist ecopoetics in the work of contemporary writers such as Stow, Gray, and Mateer, although, as we will see, Stow's spiritual values are located in Taoism.

Through the example of John Cage's Zen poetics and in light of contemporary ecopoetic theory, Peter Jaeger develops a Buddhist ecopoetics. Jaeger argues that a relationship between Zen Buddhism and ecological awareness informed Cage's work as a composer, writer, and visual artist (Jaeger 2013, 1). An understanding of Zen and a concern for environmental issues affected his writing. A Buddhist ecopoetics underlies the making of ecopoetry through principles such as dharma and ideas of ecologically engaged Buddhism, as evident in the teachings of Buddhadāsa and others. Buddhist ecopoetics draws from the inherent awareness of nature found in some teachings as a foundation for the *poiesis* (the making) of poetry. A dharmic ecology – an understanding of the natural world through Buddhist symbolic structures – translates to other places, such as Australia, through the practice of poetry. New symbols are formulated in response to the unique ecologies of the places. Having discussed the symbol, dharma and ecopoetics in this section, I now turn to the Australian poets themselves and key examples from their works in the following section.

RANDOLPH STOW'S TAOIST ANTECEDENTS

Born in Geraldton, Western Australia, Julian Randolph Stow (1935–2010) was an Australian writer who later in life emigrated to England. Some of his more acclaimed novels include *To the Islands* (1958), *Tourmaline* (1963), *The Merry-Go-Round in the Sea* (1965), and *Visitants* (1979). However, early in his career, Stow also wrote poetry. Influenced by a Buddhist grandfather, Stow's work, especially poems in the collection *A Counterfeit Silence* (1969), brings Eastern themes to bear on the perceived austerity of the Western Australian landscape, notions of human individualism, and possibilities of spiritual transcendence *in* and *through* nature. Rather than Buddhism *per se*, Stow's interests are Taoist, as evident in his Western Australian landscape poem "From The Testament of Tourmaline: Variations on Themes of the *Tao Teh Ching*" (Stow 1969, 71–75). Before providing a brief reading of Stow's poem from an ecocritical perspective, as outlined in the previous section, I must first establish the historical and philosophical tensions between Taoism and Buddhist that make such a reading plausible in the first place. The interactions between the two major Eastern religions during the last two thousand years are complex, with Taoism usually regarded as an "indigenous tradition" that began to develop more fully in China during the second century, BCE in part through the influence of Buddhist doctrines (Mollier 2008, 1). In China, the introduction of Mahāyāna Buddhism from India profoundly altered traditional Chinese religious beliefs, but also stimulated the further formation of already established Taoist principles. As an outcome of this tension, the first Taoist canon was formulated by the fifth century BCE, marking the emergence of Taoism as an institutionalized and widely practiced religion in China. Indeed, the Buddhist sūtras preached by Buddha and the Taoist scriptures revealed by Laozi both condemned the moral decadence of humanity and shared an eschatological focus (Mollier 2008, 3). Particular schools of Buddhism, such as Ch'an (known as Zen in Japan), hold many beliefs in common with the principles of Taoism. Christine Mollier's study *Buddhism and Taoism Face to Face* (2008) provides further detail about the intricate connections between Taoism and Buddhism in medieval China. However, it is sufficient to remark, without going into greater depth here, that the long-standing relationship between Taoism and Buddhism underlies my selection of Stow's Taoist landscape verse as the first in this series of Australian poems.

Of immediate note in Stow's poem are the gaps in the numbering of its stanzas. We find stanza I, followed by stanzas IV through VIII, then X, XII, XVI and XXV and finally the concluding stanza XL. This temporal structure, with its lacunae, symbolizes the extended silences of the desert landscape about which Stow writes. The lacunae also represent the difference between Eastern and Western temporal orders, the latter seeking the strict linear progression of time and the logical culmination of events without gaps or absences. Stanza I opens with the tension between naming, language, landscape, silence, and emptiness. A Western epistemology would seek to name and vocalize the land that "breaks into beauties" (l. 1) (consider the profuse flowering of indigenous plants during September and October in Western Australia and the scientific names that demarcate the species there). However, "Tao is a sound in time for a timeless silence" (ll. 4–5). Analogous to the mirror between dharma and nature discussed previously, "the land and Tao are one. / In the love of the land, I worship the manifest Tao" (ll. 8–9). Stow here invokes the historical issue of the perceived emptiness of the Australian landscape, which his forebears knew erroneously as *terra nullius*, a space thought to be devoid of life and owned by no one. In fact, some of the most biodiverse habitats on earth exist in Western Australia (consider the South-West Botanical Province; chapters 3 and 4) and some of the oldest human cultures. But the sense of space – the general lack of large trees, the flatness of the topography for the most part, and the far reach of the eye across the terrain and into the cerulean sky – is profound for most human observers. Hence, rather than a Western metaphysics of space and time, Stow adopts an Eastern one in order to express that "the land's roots lie in emptiness" (ll. 10–11). The notions of spatiality, temporality, and ephemerality reflect Robert Gray's, as discussed in the following section.

The poem pivots against these kinds of contrasts: that love of land is attributed to joy while Tao is "passionless, unspoken" (l. 7), that emptiness can exist alongside and within fullness. Stow expresses these complementarities through a Taoist symbologic structure, including repetition of the word "Tao" itself. For Stow, the animating force of the landscape is the Tao, comparable perhaps to the omnipresent Creation Beings of Aboriginal people, including the

Waugle of South-West Australia, a giant serpent responsible for the watercourses and geomorphic patterns of the land. Stow begins to develop his Taoist metaphysics of the land in stanza IV:

> The spaces between the stars
> are filled with Tao.
> Tao wells up
> Like warm artesian waters. (ll. 12–15)

Tao is at once emptiness (the spaces) and fullness (artesian waters). This instigates a reckoning with Australian space and place – historically misinterpreted by colonists and visitors as only emptiness, as *terra nullius* – made possible through the Tao. It is chthonic, encompasses polarities and precedes Christian notions of God as well as the colonial miscues towards the land: "Where is the source of it? / Before God is, was Tao" (ll. 20–21). As the poem progresses, the concrete symbols of the smith, forge, and bellows signify the creation of the patterns of the place; there is a movement between tangible symbols and metaphysical concepts with the land serving as the medium for the poet's contemplation. And all throughout is the unyielding principle of change as "world has no life but transformation" (stanza vi, l. 43). A reflection on the limits of the Western temporal order carries over to the nature of the senses and, hence, the Tao is phenomenological in character, as expressed lyrically in stanza XII:

> The colours of time blind the eye to timeless colours.
> The music of time dulls the ear to timeless music.
> The flavours of time spoil the palate for timeless flavours.
> The diversions of time dull, blind and spoil the mind

From abstraction, the poem returns to ecological particularities, as we will also see in Gray's poem, in stanza XVI as "the long roots of myall / mine the red country / for water, for silence." *Myall* most likely refers to an acacia, a plant genus known for tolerance to droughts and its ability to fix atmospheric nitrogen into the soil. The "blossoming myall" symbolizes the seeking of silence and the rhythm of the land, while the color red represents the essence, the Tao in stanza XL:

> The red land risen from the ocean
> erodes, returns; the river runs earth-red,
> staining the open sea

In sum, Stow's poem presents an interpretation of the Western Australian environment through Eastern symbolisms (Taoist, rather than Buddhist, in this instance) that intersect with Aboriginal ideas of country. Running through the poem are themes of timelessness, silence, and emptiness. These themes are constructed in positive, rather than pejorative, terms, through the Taoist lens Stow develops. A postcolonial, posthumanist ecopoetics is achieved through Eastern symbolisms, enabling Stow to observe the character of the land itself while rejecting, albeit more tacitly, its historical misconstruction as a terra nullius, a dead space lacking life and energy.

ROBERT GRAY'S CRITICAL BUDDHISM

Born in 1945 in Port Macquarie, New South Wales, Robert Gray is a well-regarded Australian writer and critic. He was influenced by East Asian Buddhism early in his career, especially as a means to respond to the Australian landscape and to conceptualize humanity as an agent within nature, rather than an entity apart from it (Langford 2012). Hart describes Gray as "not the first Australian poet to look to Asia with religion in mind, or even with a didactic impulse in mind; but he was the first to engage analytically as well as lyrically with Buddhism" (Hart 2013, 279). Yet, Gray's relationship to Buddhist thought has never been straightforward and has been marked by points of philosophical divergence. In his memoir, *The Land I Came Through Last* (Gray 2008), he alludes to his initially conflicted reaction to Buddhist doctrine, particularly what he understood as the separation of mind and body, which I will quote at length here:

> I found that, despite trying, I could not accept orthodox Buddhism, because it was built around the belief that we are reborn, even though we have no unchanging soul, and must continue to be, always suffering, while motivated by desire. Buddhism required the separateness of mind and body, since something was passed from life to life, but the evidence of research showed that any mental phenomena – volitions, memory – coexist exactly with activities of the brain. (Gray 2008, 298).

Gray's account of his early interest in Zen also indicates his alignment with a "critical" Buddhism, which he considered to embrace the natural world as a positive, sympathetic phenomenon. As such, Gray's work deploys Buddhist symbolism in developing an aesthetics of Australian landscape, but also an ethics of nature that carefully considers the place of humanity and human experience.

As further indicated by his recollections, a transformative moment for Gray was finding copies of Alan Watts' books *The Way of Zen* and *Psychotherapy East and West* and their orientation to critical Buddhism, the aim of which is to attain a degree of absorption in the world or "self-forgetfulness through intense involvement" (Gray 2008, 219). Gray's philosophical association with Zen Buddhism anticipates, in a way, the later formalization of the Critical Buddhism movement within Japanese scholarship, evolving in particular from the writings of Matsumoto Shirō and Hakamaya Noriaki who argue that Buddhism is inherently based on the principle and practice of criticism. Accordingly, for these scholars, Critical Buddhism differs to Topical Buddhism, the former interested in rhetoric and ontology and the latter in logic and epistemology (Lin 1997). However, it should be held in mind that Gray's critical Buddhism ("c" in the lower case to distinguish Gray's thinking from Shirō and Noriaki's) is slightly different and more specific to his maturation as an Australian poet writing in Australian settings. Moreover, Gray's Buddhist ecopoetics attends to the interpenetration of matter and spirit, of nature and humanity, through protracted, sense-rich engagement with the local environment – coastal New South Wales and other places, including the Western Australian landscape of Stow's work.

Gray's dialogue with Buddhist ideas is prominent throughout most of his collections, including *Certain Things* (1993) and *Cumulus* (2012), in both existential and ecological terms. However, the poem I will focus on briefly is titled "Dharma Vehicle" from his earlier collection *Grass Script* (1979). This long, seven-part poem employs the principle of dharma throughout as a means to address environmental and place-based themes. Given the range of interpretations of dharma by different Buddhist scholars, as indicated earlier in this chapter, it is instructive to consider how Gray himself conceptualizes the term. In a different poem, "To the Master, Dogen Zenjii" (Gray 2012, 24–27) defines *dharma* as the transience of the world in the following quote from Dogen Zenjii (1200–1253), the Zen Buddhist teacher of Kyoto, Japan: "It is this world / of the dharmas (the momentary events) / that is the Diamond" (24, ll. 16–18). Indeed, following Dogen's idea, Gray's "Dharma Vehicle" begins with the momentary events of perception in a characteristically Australian setting:

> Camping at a fibro shack
> fishermen use—
> swept with ti-tree branches, and washed down
> with kerosene tins of
> tank water

Like banners raised,
all these eucalyptus saplings—
the straight trees

The scene, allusions, and diction are quintessentially of Australia. A "fibro shack"
is a colloquialism for a small asbestos structure, consisting of fibrous cement
sheets. These dwellings, of various sizes and styles, are common throughout the
built environment of Australia, including in coastal areas. "Ti-tree" refers to a
species of Melaleuca, plants in the myrtle family and widely distributed in the
country. In addition to ti-trees, eucalyptus trees populate the landscape in
profusion and also are referenced in the poem. Moreover, "tank water" refers to
the rain water from collection tanks found attached to many Australian homes, as
a means to capture a precious resource in a drought-prone landscape. We find in
part 1 of the poem a strong sense of movement, transience, and poiesis as "leaves
here / are shaken all the time" (ll. 14–15) and "my bed / a pile of cut fern" (ll. 18–
19). Hart (2013, 277) observes how the landscape depicted in the poem's opening
is "palpably Australian" until the paperbarks, another plant species typical of
many habitats, are likened to "incense-smoke," thereby marking a transition to
Asian symbolism. The poem encapsulates Buddhadasa's four-fold definition of
dharma as "nature itself; the laws of nature; a person's duty to act [...] and the
benefits to be derived" (1989, 128), as well as Dogen's idea of momentary events.
However, this expression of dharma is achieved in Gray's poem through
"palpably Australian" icons: paperbarks, ti-trees, fibro shacks, water tanks. Thus
a dialogue between Eastern and Western symbolisms is struck.

An emphasis on movement and temporality, through the stories of Gautama
and Heraclitus, continues in the poem's third part with an assertion that the
human soul is the same as nature's (Gray 1978, 39). But Gray's Buddhist
ecopoetics is not merely about distantly observing the transience of the natural
world through moments of perception and insight. The oscillation between the
minutiae of the Australian landscape and the contemplation of Buddhist tenets
and stories imbricates human agency, as well as the failures of human actions in
nature. In terms of human agency, consider the following compelling excerpt:

So that these transient things, themselves, are what is Absolute;
these things
beneath the hand, and before the eye—
the wattle

lying on the wooden trestle,
pencils, some crockery,
books and papers, a river stone,
the dead flies and cobwebs
in the rusty gauze. (part 3, 40)

For Gray, the transient things are dharma, presented to perception via sight (a sense long associated with distance, space, intellection, and rational inquiry) and touch (a sense of intimacy, proximity, and subjective knowledge). The wattle (the common name for Australian acacias) along with the implements of the writer's life and practice constititute, in Hart's (2013) words, the "field of dharma." Here we clearly find the transmission of Buddhist ideas to a new medium: the Australian environment and, more importantly, the intersection of human agency and place ecologies. Furthermore, in part 4 of the poem, Gray's critical Buddhism is insinuated in observations of "the paddocks below / amongst innumerable dead trees" (41). Banana plantations have resulted in the cutting back of the bush as "the early sky, so light / has a feeling of / the first day up again after illness" (41). Hence, the natural world observed by the poet is neither the object of visual beauty solely (but rather also of the tactile senses too) nor an Edenic field unmarked by human action (but rather one bearing negative impacts). Although the tension is palpable, this anti-pastoral hint in Gray's poem sits comfortably alongside his Buddhist ecopoetics, especially in relation to his acknowledged critical Buddhism.

JOHN MATEER'S SERPENTINE SYMBOLISM

The final poet I will discuss in this abbreviated overview of Buddhist Australian poetry and its ecocritical dimensions is John Mateer. Born in 1971 in South Africa, Mateer is an Australian poet and author with a long-stranding interest in the Western Australian landscape, but who has also travelled and lived in Southeast Asia. The subject matter of his short poem "The Serpentine Monastery" in his collection *The West* (2010) is Bodhinyana, a Theravadin Buddhist monastery in the town of Serpentine, located south of Perth, WA. Bodhinyana is in the Thai Forest tradition of Buddhist monasticism in which practitioners form a close relationship to the forest and wilderness settings of the community. The dominant symbol in the poem is the snake, both in Buddhist and Aboriginal Australian contexts. For instance, *nāga* is a Sanskrit word for a class of beings that take the shape of a king cobra. The tradition of a great snake is found in all Asian Buddhist countries. In Tibetan Buddhism the *nāga* is related to the *klu*, dwelling in surface and underground water bodies and often protecting treasure. Another notable *nāga* in Buddhism is Mucalinda, a serpentine being who protected Buddha as he meditated under the Bodhi Tree and was exposed to darkness and rain. Representations of Mucalinda are common in the Buddhist art of Laos. Similarly, for the Nyoongar people of the South-West of Western Australia, the *Wagyl* (or *Waugl*, *Waugal*, or *Waagal*) is a snakelike creature governing the surface and underground water of the region, including the major river courses of the Swan and Canning Rivers (chapters 3 and 4). The Darling Scarp, the hill features surrounding Perth, are known as the body of the *Wagyl* and the abundant rock outcrops in the Wheatbelt area east of Perth are regarded as his droppings. In a periodically arid landscape, respecting the *Wagyl* has helped to ensure the continuity of water resources, in the form of lakes, rivers, springs, wells, and water holes.

Mateer's poem addresses the complexities of this intersection, between the serpentine symbologies of the East and West. The poem opens with an observation of almsgiving: a spoonful of rice and the act of bowing to the resident monks. As with Gray's oscillation between the abstract and the concrete in "Dharma Vehicle," Mateer's work also shifts from contemplating the principle of "non-proliferation" (134, l. 4) to experiencing the palpable reality of the Australian bush. He notes the "ball-bearing gravel," "the parrotbush" (a local plant species in the *Dryandra* genus) and the "contorted banksias" (ll. 5–6), all "impersonal" to him as the monastery itself, the doctrine with which he grapples, and the landscape to which he returns. The characterization of the Western Australian bush as impersonal or unwelcoming (*i.e.* as prickly as its plants) is nothing new (chapters 7 and 8). As *Posthuman Plants* has already explored, the Australian historical record abounds with these sorts of references. Although the local environment becomes a canvas for the poet's unease, there is still an attentiveness to nature, to dharma, evident here. Western Australia is, by all accounts, an isolated state: its capital city of Perth is several thousand kilometers from the nearest metropolitan area of Adelaide, South Australia. Yet, in this sequestered island on the western edge of the country, this "refuge from the new economies" (l. 7), the encounter between Eastern ideas and Western ecologies goes on. The snake (indeed, the final words of the poem), therefore, symbolizes the Indigenous ecologies of the place (including its Dreaming stories and the ancient flora surrounding the monastery) yet the resistance of the place (and its people) to the deeper penetration of traditional Buddhist doctrines into a field of dharma. Simply put, the formulation of a quintessentially Australian dharma, for Mateer, is not seamless or harmonious, but involves critical concern for the contradictions he sees as inherent to some Buddhist principles, especially when enacted in other contexts.

CONCLUSION: BUDDHIST ECOPOETICS IN AUSTRALIA

Through the works of Stow, Gray, and Mateer, this chapter has argued that Buddhist Australian poetry characteristically involves local ecologies and Australian forms of life, such as paperbarks, banksias, ti-trees, and others. Endemic flora, fauna, and ecologies become symbols for Buddhist principles as well as subjects of dharmic enquiry. An ecocritical, posthumanist perspective on Buddhist Australian poetry necessitates the consideration of the ecologies of Australian places in relation to the enactment or investigation of Buddhist principles. Thus, living plants and animals are not only ecological beings, but are imbued with dharmic meaning, as the poets attempt to know nature, its principles, and the benefits of acting in accordance with natural laws. More importantly, nature in these poems is not a distant object of contemplation or observation, but rather a living conversant in everyday life, with which the poet is immersed in the world. The botanical and human worlds are permeable. Notably, we find multisensorial expressions of engagement with nature (as dharma) as part of an Australian Buddhist ecopoetics. I suggest that further research into this subject should consider additional poems by these writers and others, such as Judith Beveridge, whose exposure to Buddhist symbols has influenced their poetic practices of landscape.

PART V

Plants & Senses

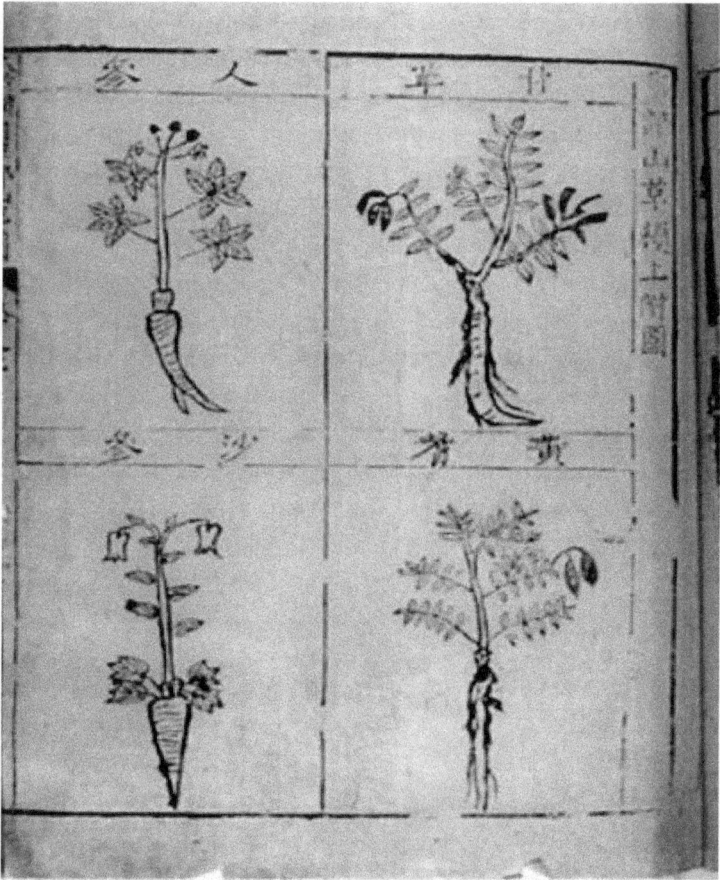

《本草纲目》明、清善本（金陵本书影 1-149 中研院）
Illustration from *Ben Cao Gang Mu* compiled by Li Shizhen (1590 AD)

CHAPTER 10

In the Key of Green? The Silent Voices of Plants in the Poetry of Glück and Bletsoe

INTRODUCTION: THE CACTUS AND THE HOYA

> Everything which is, whether animal or vegetable, is full of the expression of that use for which it is designed, as of its own existence … Let [us] respect the smallest blade which grows, and permit it to speak for itself. Then may there be poetry, which may not be written perhaps, but which may be felt as a part of our being. –Sampson Reed (1859, 55)

Since this plaintive appeal by Reed, allowing the "smallest blade" (or, prickliest spine or loveliest heart-shaped leaf) to speak has become a technological preoccupation for some (see chapters 5 and 6). Let us begin with a typical example. Cactus Acoustics aims to allow saguaro cacti to vocalize (Oskin 2013). We might imagine the voice of the burly saguaro as gruff and slightly imposing. Growing to considerable proportions – up to five stories high, eight tons in weight, and over a hundred years in age – *Carnegie gigantean* is endemic to the Sonora Desert. As an adaptation to aridity, the cactus, at times, becomes an oversized sponge, absorbing hundreds of gallons of water a day during seasonal deluges (Dimmitt 2000). Using an acoustic detector, researchers hope to correlate the sounds produced by the Sonoran giants to water fluctuations, temperature extremes, and ultraviolet exposure. In other words, they intend to elicit the environmental vocabulary of the saguaro: "I'm cold versus I'm really thirsty" (quoted in Oskin 2013, "How to Listen"). The result could be a device allowing plants to express their needs – endowing even the most hopeless plant minder with a green thumb.

The idea of plant-voice takes a radical turn in this example, leading farther from a poetry (chapters 7, 8, and 9) (but closer to a technics) of being. Let us consider another instance of the "human desire for universal communicability" (Marder 2016). The proprietor of a Japanese café affixed sensors to a *Hoya kerrii*, known as sweetheart plant or lucky-heart for its cordate leaves. The equipment detects the bioelectric current of the sweetheart in response to its setting (including human movements in the café), and then renders the signals into

Japanese words. Affectionately named, Mr. Green composes a daily blog, including his observations of the weather: "Today was a sunny day. I was able to sunbathe a lot" (quoted in Barras 2008). Internet users activate a lamp for his pleasure: "Being able to receive full light from the rear is delightful!" (paraphrased from Cordis 2008). Seemingly novel, the hoya project owes its lineage to digital works that have claimed to allow plants to speak, fly, and express their creativity (see chapter 6). Yet, the voices in these examples have distinctly (and perhaps eerily) human tenors. How might we permit a cactus and lucky-heart to speak for themselves without "objectifying them or, at best, speaking for them, in their defense, if not in their place" (Marder 2013a, 186). How might poetry assist us in doing so?

THE (IM)POSSIBILITY OF PLANT-VOICE?

Just as the lucky-heart is delighted by the lamp's warmth, so too are humans enthralled by speaking plants, even those with artificial voice boxes. As listening to plants becomes an evermore-mediated activity, new mechanisms are purported to bridge the human-vegetal chiasm while also engaging our imaginations. Such innovations are thought to promote interspecies dialogue by enabling plants, for example, to tell us when to water or feed them (Faludi 2013). The problem is that, despite their endearing intents, these interventions are built upon an audiocentric logos privileging a narrow, monologic notion of voice. According to this logos, a plant voice must speak if it is to be heard; if it does not speak, it is mute, an utterance of no consequence, with neither register nor agency; and, furthermore, plant-speak requires technical prosthesis. The wild saguaro, the café sweetheart, and the household philodendron might eventually be able to communicate to us, in clumsy or sophisticated diction, but only by approximating human speaking, that is, by "ventriloquizing" (Marder 2016). As Michael Marder further argues, "the assumption that to have a language is to be able to speak is both erroneous and unethical [...] it ties linguistic phenomena to the voice, which only humans possess" (Marder 2016).

The chortling hoya and the gruff saguaro are designed to fulfill a desire for communion with the botanical world – one that is not normally within the scope of everyday interactions. However, these desires are met only partially in a manner that approaches plant puppetry – in which voice is the outcome of a kind of cyborgian vegetal subjectivity, of the humanization of the plant, of the making of the vegetal in the image of the animal. In asking "will the Saguaro give a scream when he is thirsty and make a different noise when he is too cold" (Wardell and Rowe 2010), the botanical imagineers of today might very well be ruling out the silent voices and embodied expressions that constitute the language of plants in all its complexity. Do these interventions disclose plant-voices that have always been, or do they impose (or construct) voices that have never been – and are perhaps not meant to be, if indeed it is only humans who possess voice? Is the possibility of plant-speak valid through a scientific basis and culturally acceptable only through technologized examples like the saguaro and ventriloquized ones like Mr. Green? How might it be possible to think of voice otherwise – as non-verbal, bodily, and ecological articulation, and as an ontological concern rather than an auditory phenomenon – and how might plants particularly help us in doing so? How might poetry be a medium for hearing and listening to plants – where language (rather than electric sensors and algorithms) becomes a shared and porous interface between the speaking us and the silent plant other?

This chapter will pursue these challenging questions in developing a concept of plant-voice that resists, through language, the technological mediation of the cactus and the hoya. It will attempt to respond to Jennifer Peeples and Stephen Depoe's fundamental questions: "What is nature's voice? Does it 'speak'? If so, how? To whom? Can humans attend to the voice of nature" (Peeples and Depoe 2014, 9). It will follow Reed's cue from over one-hundred-and-fifty years ago by proposing that, within critical plant studies, it is timely to reconsider the intricacies and possibilities of plant-voice. The emphasis will be on language, not as human enunciation or a symbolic system but as "the house of Being" (Heidegger 1982, 63) or, building upon the work of Eric King Watts (2001, 179), as "an original impulse of being." Divided into two parts (Theorizing Plant-Voice and Poeticizing Plant-Voice), the chapter will develop a posthumanist and ecological (*i.e.,* relational and material) concept of plant-voice, firstly, through a critical review of ecocriticism and environmental communications studies and, secondly, through the contemporary poetry of Louise Glück and Elisabeth Bletsoe.

Glück and Bletsoe engage substantively with plant-voice, intersecting as a result with issues of vegetal heteroglossia, embodiment, and ethics. As such, plant-voice is the material elocution of the vegetal in its milieu, rather than a symbol of something outside itself. In their poetry we witness a movement from plant-voice as metaphor or literary maneuver to one intoned ecologically, that is, as an outcome of a plant's interactions with other beings, its material environment, and human culture. Glück's *The Wild Iris* (1992) is a botanically inflected collection that received the Pulitzer Prize. The work addresses themes of identity, domesticity, and plant-human communication in which the poet, adopting the perspective of flowers, grapples with the heterogeneity of voice. Without speaking from the point-of-view of plants, Bletsoe's *Pharmacopœia* (2010) integrates knowledge of herbal medicines and botanical folklore as expressions of the historical plant-voice articulated in the present. I will argue that the poetic invocation of plant-voice should not always be read pejoratively as pathetic fallacy. The examples will show that plant-voice in poetry needs to be grounded in the material realities and sensory expressions of the plant in relation to its environment (including the human environment) and the creative intents of the writer. That plant-voice depends on human desire (in this instance, toward poetry) affirms, rather than undermines, the extent of our interdependence with vegetal life and the need for ethical voicing.

PART I: THEORIZING PLANT-VOICE

First appearing in English in the late thirteenth century, the word voice derives from the Old French *voiz* for "speech or word" and Latin *vocem* for "sound, utterance, or cry" (Harper 2014). As a noun, voice can refer to the expression of feeling (as in "the voice of the people") or, as a verb, the act of expressing ("to voice an opinion"). However, voice normally denotes the sounds made by the human larynx, mouth, tongue, and lips to communicate in distinct tones and accents, making individuals known to other beings. To suggest that plants have voices might seem absurd or erroneous; anatomical sense tells us that plants cannot vocalize as humans do. For plant-voice to be possible, then, we must think about voice differently, while refusing its purely metaphorical association with the vegetal. This dilemma is evident in science, where plant-voice is a contentious and provocative figuration. The field of bioacoustics empirically suggests that plants have a kind of voice yet the term is applied cautiously as a figure of speech and tellingly in scare quotes in relation to new scientific findings about plant communication (Vieira, Gagliano, and Ryan 2016, Trewavas 2014). Phytoacoustic research shows that plants emit unique sound signatures, enabling decision-making and survival through communication with their habitats (Gagliano 2012). The ecological function of sound by extension implies attributes of agency and, arguably, intelligence in a form of life that has been constructed as the antithesis of the animal – as mute, passive, and largely devoid of cognitive powers (Ryan 2013, chapter 6).

Voice has also been a complex issue in philosophical and literary studies of nature. In these fields, plant-voice has largely been assumed, shelved, or sidestepped as improbable – a fracture in an otherwise tenable thesis about the natural world, an imperfection in the narrative of a sentimental writer, a transgression against rational discourse, or an unfortunate flight of fancy. After all, nature can't speak as we do. However, paradoxically, since the growth of the field in the 1990s, listening to nature's voice has been seen as central to the aims of ecocriticism, as Michael McDowell explicates: "Beginning with the idea that all entities in the great web of nature deserve recognition and a voice, an ecological literary criticism might explore how authors have represented the interaction of both the human and nonhuman voices in the landscape" (McDowell 1996, 372). Considering these movements toward, biases against, and inconsistencies with the concept, this discussion will stir the sleeping dragon of nature's voice as a tacit and fundamentally unresolved issue within ecological literary criticism.

Nineteenth-century British art critic John Ruskin coined *pathetic fallacy* to describe the attribution of feeling, emotion, and sentience to so-called inanimate nature. For Ruskin, this involves a "morbid state of mind, and comparatively [...] a weak one" (Ruskin 1998, 72). A botanical case in point from "Spring" (c. 1850) by American writer Oliver Wendell Holmes is leveraged. The poem depicts a crocus imaginatively and sympathetically, but in androgenic terms: "The spendthrift crocus, bursting through the mould / Naked a shivering, with his cup of gold" (Holmes 1870, 64). The hard facts of ecology reveal Holmes' lines to be "very beautiful, and yet very untrue. The crocus is not a spendthrift, but a hardy plant; its yellow is not gold, but saffron" (Ruskin 1998, 64). The distinction between human and nonhuman – and between intellect and imagination – is drawn sharply by Ruskin. (The crocus is not naked, but rather in its natural state; it does not humanly shiver, in response to cold or fear.) The protracted sentimentality of the poem would surely have caused Ruskin great anguish, from "her clustering curls the hyacinth displays" to "the robin, jerking his spasmodic throat" (Holmes 1870, 225-227). (Retorts Ruskin: The hyacinth hasn't curls, but petals and sepals; the robin's voice is not spasmodic, but how it attracts a mate.) Meanwhile, Holmes lamented his own powerlessness to hear nature speaking. Despite his figurative skill, he felt the "chains" of science (and ratiocination) ultimately separating him from the voices that stir and beckon him during this season: "Why dream I here within these caging walls / Deaf to her voice, while blooming Nature calls" (Holmes 1870, 227). Thus we find a polarizing tension between the seeking of voice in Holmes' poetry and the refusal of voice in Ruskin's criticism.

How far has the principle of nature's voice come since Ruskin and Holmes? Not very. Firstly, in ecocriticism, it remains overshadowed by the specter of the pathetic fallacy and the residues of modernist anti-sentimentality (involving an aversion toward expressions of emotion in poetry) (Greenberg 2011, 11–16). McDowell even concedes that pathetic fallacy, as the "crediting of natural objects with human qualities," is inevitable and "something to acknowledge and celebrate, not to condemn," assuming that this is inherently negative and failing to tell us how to celebrate it (McDowell 1996, 373). Secondly, closely associated with English Romanticism and American Transcendentalism, nature's voice is usually conceptualized in sonic terms, as the speaking position of nonhuman beings. Speaking is vocalizing; if something hasn't voice, then it cannot be said to speak. Thirdly, to attribute voice to nature is to elicit its moral consideration. If plants have voice, then they should be included within an ethical domain. Conversely if they lack address they are more conveniently relegated to the background (Marder 2013a).

These concerns (of tradition, modality, and ethics) play out as ambivalence toward voice in mainstream ecocritical scholarship. Greg Garrard (2004, 47) describes the "post-Romantic problem" of voice as "necessarily human and 'reflective' and yet almost naively open to the natural 'other'." The silent voice of nature, for Garrard, is an oxymoron; never lacking sonic register, voice is auditory expression. This audiocentric bias is raised in relation to Percy Shelley's poem "Mont Blanc." The mountain is endowed by the poet with voice and, by extension, agency: "Thou hast a voice, great Mountain, to repeal / Large codes of fraud and woe" (Shelley quoted in Garrard 2004, 65). A Heideggerian "letting beings be" is posited as a solution to the inescapable anthropocentrism of giving voice – as a means of releasing voice from sentimental human states of mind and desire (Garrard 2004, 47). How successful this decoupling of voice from reflection is we cannot be sure, as the possibility is presented in passing. However we can be more certain that, for Garrard, following ecocritic Jonathan Bate, the viability of nature's voice as a concept is contingent on a dwelling in language, rather than constructing language as symbolic or referential. The voice of the mountain is less about *language* and more about the language of things (Benjamin 1996). However promising, this line of argument is only faintly sketched.

The presumption that nature's voice is out of the reach of Western consciousness is principal in Bate's *The Song of the Earth*. He posits that the "inheritors of the Enlightenment's instrumental view of nature" can only conceptualize voice as a metaphor, not able to understand dialogue with plants as a legitimate form of discourse (Bate 2000, 165). Bate admits that this could differ for Aboriginal Australian and other Indigenous readers for whom the land sings and is impregnated with meanings and voices – where a dialogic tradition between humans and nonhumans exists traditionally in everyday awareness (Graham 2008, Rose 1996). However, rather than confronting the stigma of pathetic fallacy and opening up possibilities for nature speaking (and, for that matter, nature writing), a series of speculations about John Clare's "The Lamentation of Round-Oak Waters" instead implies the implausibility of voice. Here, a stream vocally protests its environmental conditions. We might reflect back to the sentient saguaro ventriloquizing through an acoustic detector. But there is a difference: the stream's habitat has been fragmented by the British enclosure system of the nineteenth century. Although provocative, his queries sidestep an unsettled (and unsettling) issue in ecocriticism:

> Is the voice of Round Oak Waters to be understood only as a metaphor, a traditional poetic figuration of the genius loci, or 'an extreme use of the pathetic fallacy'? Or can we conceive the possibility that a brook might really speak, a piece of land might really feel pain? (Bate 2000, 165)

To compound this ambivalence, Bate asks if attributing the ability to speak to nonhumans corresponds to their capacity for pain. Is there a potentially dangerous correlation between speaking, agency, and moral consideration? In other words, if the boundary between the animal and the plant is blurred, then what will the consequences be for societies that now callously exploit the botanical kingdom as mute material and voiceless resource? Unfortunately, for plants and us, these questions are left unanswered.

Plant-Voice as Potentiality

For ecocriticism, environmental communications, and critical plant studies, the voice of the vegetal should be a core theoretical concern. It is productive for scholars in these fields to re-imagine the potential of nature's voice in order to understand its relationship to literary, cultural, and scientific works, rather than dismissing it as anthropocentric or questioning it open-endedly. This shift involves being aware of the preconceptions of theorists for whom voice is both figurative and contingent on human subjectivity. For instance, Christopher Manes (1996, 25) argues that "attending to ecological knowledge means metaphorically relearning 'the language of birds' – the passions, pains and cryptic intents of the other biological communities that surround us and silently interpenetrate our existence." For Manes, voice is symbolic, enigmatic, and forgotten; relearning involves deciphering, in which nature is deconstructed like a linguistic code or structurally analyzed like a text. As another example, in *The Natural Contract*,

Michel Serres (1995, 48) observes "the bond that allows our language to communicate with mute, passive, obscure things – things that, because of our excesses, are recovering voice, presence, activity, light." Although ethically inflected, this statement reveals a tacit anthropocentrism in which "mute" things have voice only if we choose to correct our "excesses" and grant them a speaking position.

The potentiality of voice is also constrained by the stigma of nostalgia – that accepting voice is an undesirable recoiling to a prediscursive idyll or an abyssal leap forward: "But where is the voice of nature calling us? Back to the pre-modern age? Or forward to a saner future" (Bate 2000, 36)? This sense of polarized temporality (past or future) is compounded by a sonic bias (in which the auditory dominates). As a further example, Lawrence Buell (1995, 152) regards the voice of Walden Woods as a synecdoche rather than an actual attribute of the forest, notwithstanding Henry David Thoreau's ostensible belief otherwise. For Thoreau, voice was something real though inaudible, stating in *Walden* (a seminal text in ecocriticism and nature writing, originally published in 1854; see chapter 12) that "the language that all things and events speak without metaphor" (Thoreau 2004, 119). His concept of *effluence* is illustrative. It concerns the olfactory register of fruits and flowers – an expansion through sensory awareness of nature of the bounds of voice and language. Through effluence, the perceiver and the perceived are more intimately connected; the naturalist becomes receptive to the diverse expressions (*i.e.* languages) of species (Homestead 2014, 197) (chapter 12).

Environmental communications theory has intensively focused on the relationship between voice and nature, offering the potential to enhance the ecocritical debate (Carbaugh 2014, Watts 2001, Watts 2013). For these scholars, nature's voice is not an intractable issue to be skirted but, to the contrary, one that contributes to a critical re-evaluation of the very premises of voice itself. Peeples and Depoe (2014, 9) argue that "these questions stretch the theoretical and material dimensions of how we understand the relationship between voice and the environment." The voice of the crocus or cactus necessitates rethinking language and communication, in order to postulate an affective theory of voice. As affect, voice is extended to linguistic and nonlinguistic acts alike, leading to ethical implications for the rhetor and the receiver. Rather than privilege voice narrowly as the attribute of a speaking subject or an outcome of linguistic discourse, Watts (2001, 180) characterizes voice as relational and never separable from the body. Voice registers being to other beings, while also communicating identity and ideology. To develop a more inclusive voice, however, we must refuse its rigid association with speaking authority and highly individuated subjectivity (Watts 2001, 192). Watts goes on to present an incisive yet poetic definition of voice as:

> the sound of affect. Voice emanates from the openings that cannot be fully closed; from the ruptures in sign systems, from the breaks in our imaginaries, from the cracks in history. It registers a powerful, some would say passionate, cluster of feelings triggered by life finding a way to announce itself. (Watts 2014, 259)

It is the particular ontologies of plants (their inaudibility, lack of address, and relative fixity in place) that further stretch the "theoretical and material" limits of voice. As Marder (2013, 32) (responding to Socrates) acknowledges, "unlike an animal, the plant has no voice (this explains its reticence), and it is incapable of spontaneously choosing its place by exercising the freedom of self-movement (which justifies its sealed character)." In contrast, I argue that the plant does have a voice, but not of the animalistic kind – a voice that perfectly corresponds to its ontology and habitus, as the plant finds a way of announcing itself. And poetry is a means of listening to and expressing plant-voice as potentiality, as "breaks in our imaginaries" and "cracks in history."

Plant-Voice as Presencing

Understanding plant-voice requires conceiving of possibilities beyond voice. Here is another key, already touched on: the linguisticity of plants becomes a matter of ethics. As Yi-Fu Tuan (quoted in Tschida 2014, 220) observes, when something is thought to fall silent, we tend to conceptualize it as dead (its presence is effectively lost). For Reed, respect for plants is the precondition for their speaking, for their having voices in the first place and expressing on behalf of themselves. The Genevan philosopher Jean-Jacques Rousseau (who was also a lifelong student of botany) claimed that human conscience is the voice of nature active within us (Bate 2000, 35). For Rousseau, neglecting to listen to nature's voice leads to existential crisis; rather than incomprehensible or mute, the voice of nature is in fact accessible to those who learn to apprehend it (Cooper 1999). But, many of us – including myself – have tried in the most respectful and sober manner (that is, without the aid of psychedelic compounds) to hear their chatter. We have tuned our ears to green things of many shapes and sizes, and in many places in the world, only to apprehend the swoosh of the wind through leaves or the thumping of our hearts in their "caging walls" (to borrow a phrase from Holmes). Compliments spontaneously uttered to flowers – "well, aren't you a beauty," as if to segue into a conversation – are invariably met with the reverberating silence of flawed yearning, like a stone rattling inside an empty steel tank (chapter 8).

To understand plant-voice as presencing, the voices of plants (their internal voices, produced by them) should be distinguished from the giving of voice to plants (their external voices, imposed upon or granted to them by us). However, there is also a middle ground that threads between these categories. This involves plants speaking for themselves, in which they express their voices in myriad ways as we present to them (and ourselves) the appropriate conditions for doing so (such as unfragmented habitat, pollinators, sunshine, respect, reciprocity; see chapter 2). This third category – which I will call an "ecology of plant-voice" – attempts to mediate the binaries of nature-culture, subject-object, human us-plant other, acting-being acted upon, and speaking-being spoken to, which limit the emergence of plant-voice as dynamic inter-relationality. Such a conception runs against the reduction of "the language of things to human language" and opens things to speaking "only to the extent that their linguisticity is not […] a matter of metaphor" (Fenves 1996, 89). It also attempts to refuse the alignment of plant-voice to the speaking position of the plant. The danger there lies in reverting to a strongly individuated, non-relational, and aurally-based conception of voice – an unachievable agenda for the plant, except with the aid of a prosthesis.

The first dimension (internal voice, which phytoacoustics points to), therefore, needs to signify something beyond audible speech or rudimentary forms of vocalization – beyond human language. An ecological plant-voice encompasses their silent presences in space and time, their sensory articulations within an *umwelt*, and their modes of signification – not mute but silent, not indecipherable but corporeal, not of the brain but of "mindlessness" (Trewavas 2002). Similarly, the second dimension (external voice, which phytopoetics points to) must involve more than our speaking for nature (consider old growth forest campaigns) or representing plants as thinking and sensing beings. While plants cannot be said to speak in modes recognizable to us (even with acoustic detectors), they do express through manifold means that at once affirm their familiarity and strangeness. Some beings use speech, but everything, including the vegetal, speaks via presence, or "voice/presence" (Saito 1998). It might, then, be possible to think of two modes of plant-voice as dialogic sides of the same grape vine; that speaking for plants in poetry, as an act informed by their ecological and material realities, is in fact more ethically inflected than not writing anything from their perspectives, or worse yet objectifying them in language. Moving beyond a concept of voice as speech or aural utterance to one of presencing, let us now consider how these ideas of plant-voice (as relational, corporeal, sensorial revealing) manifest in the poetry of Glück and Bletsoe.

PART II: POETICIZING PLANT-VOICE

Plant Rhetorics in Glück's *The Wild Iris*

Louise Glück is an American poet (b. 1943) whose work over nearly five decades beginning with the collection *Firstborn* (1968) has been described as spare with "subtle, psychological moments captured by the austerity of her diction" (Diehl 2005, 1). In confronting states of existential unease, Glück's poetry has also been characterized as stoic and confessional in its "agitated, relentless imagery and language" (Dodd 1992, 151). However, the Pulitzer Prize-winning volume *The Wild Iris* (1992) marks Glück's departure from the stable, undifferentiated writerly voice characteristic of the confessional poetics of her earlier work. Rather than a unified speaking position, *The Wild Iris* exhibits a range of subject positions in which the authorial self crosses "the border between the human and the not human, as under construction, and in a state of becoming" (Morris 2006, 200). The work has been called a "polyphonic theater," "heteroglossic text," "prayer sequence" (Morris 2006, 191), "radically heteroglot volume" (Gordon 2002, 228), and "lieder cycle [...] written in the language of flowers" (Helen Vendler quoted in Glück 1992, back cover). Its alternation between Matins and Vespers (used as poem titles) invokes the Hours, or the Roman Catholic morning and evening prayer cycles. Within this religious structure superimposed over a Vermont garden's seasonal cycle, *The Wild Iris* adopts three lyric categories: an omniscient God figure, a poet-gardener-supplicant, and fourteen flowering plant speakers.

The plant personae include wild and cultivated species such as the titular wild iris, trillium, snowdrops, Jacob's ladder, and hawthorn. Critics have described the interspecies dialogism of Glück's plant characters as figurative and inescapably mediated by the human voice. Accordingly, critics stress that the plant-voices are symbolic representations of God projected onto nature, metaphors of "the self when imagined as a speaking flower," and foils "for an internal conversation" in which the poet ventriloquizes plants (Morris 2006, 221, 230). Similarly, others observe the human-like voices of the flowers, conceding flatly that "the human writer has no other voice to give them" (Gregerson 2001, 117). While it is irrefutable that the text constructs flower voices in human terms, a reading of Glück's engagement with ecological plant-voice will elicit nuances not possible in these analyses. The speaking flowers vocalize in English diction, of course, but also articulate spatially and sensuously as an expression of their ecological situatedness. Their speaking is plant-voice as "life finding a way to announce itself" (Watts 2014, 259) in which the poet and her audience become witnesses to their language and facilitators of its textualization (where the flowers still speak to readers like me). Glück's plant-voice is underpinned by an appreciation of ecological cycles, relationality between beings, and "the shared materiality of the earth body and the personal, particularly the female, body" (Gordon 2002, 222).

The first poem in the volume, "The Wild Iris," is written wholly from the plant's perspective and is redolent of the Greek myth of the vegetation goddess and ruler of the underworld, Persephone (Glück 1992, 1). The iris flowers and leaves die back then regenerate from the bulb, offering a compelling symbol for natural cycles of life and death and signifying metaphysical reflection on nature (Morris 2006, 200). However, to read the voice of the iris figuratively shuns the material realities of the work, as well as the close association between voice, consciousness, and spatial articulation it develops. The iris' direct address to the poet-gardener discloses its chthonic memory, reaching into the depths of time and earth and exceeding the limited human capacity to remember: "Hear me out: that which you call death / I remember" (Glück 1992, 1, ll. 3–4). Moreover, the iris (specifically its rhizome) is self-aware, enduring seasonal interment "as consciousness / buried in the dark earth" (Glück 1992, 1, ll. 9–10). Glück's poetic assertion might not be far from the empirical reality of plant cognition. Scientific research dating back to Darwin in the nineteenth century has suggested controversially that the brain of the plant is located in its root tips (Barlow 2006).

Through the iris' voice, its consciousness, spatial awareness, sound perception, and a poietic sense of time become predominant in spare lines such as the following: "Overhead, noises, branches of the pine shifting" (Glück 1992, 1, l. 5). Rather than an auditory phenomenon, plant-voice here is a vinculum between the different articulations of the iris announcing its presence to other beings, including the poet-gardener-supplicant and the reader. The concluding stanzas posit plant-voice as the sensuous, non-verbal expression of the flower negotiating its subterranean and above-ground milieux:

> You who do not remember
> passage from the other world
> I tell you I could speak again: whatever
> returns from oblivion returns

to find a voice:

from the center of my life came
a great fountain, deep blue
shadows on azure seawater. (Glück 1992, 1, ll. 16–23)

The iris' mode of address approximates Watts description of voice as "the enunciation and acknowledgement of the obligations and anxieties of living in community with others" (Watts 2001, 180). The iris-voice resonates with tactile ("great fountain"), visual ("deep blue"), and sentient ("returns from oblivion") expressions, as the flower emerges from its underground dormancy not as an individual subject, but in community with "the stiff earth," "birds darting in low shrubs," and other living beings and material things.

Throughout *The Wild Iris*, plant speakers make themselves known spatially, while the poet observes the workings of ecological voice in her garden. A few more examples from the volume will show the polyvalence of plant-voice. Empathic feelings between plants and humans are part of the text's voicing. The purple flower of "The Jacob's Ladder" addresses the grieving poet, presumably lovesick, from outside her bedroom window. The self-aware flower enunciates itself gesturally in a tone of longing evident in its "naked stem / reaching the porch window" (Glück 1992, 24, ll. 10–11). The flower's language, physically pronounced, exudes empathy for the woman and shares with her a desire for transcendence from their earth: "Never / to leave the world! Is this / not what your tears mean" (Glück 1992, 24, ll. 13–15)? In addition to the flower-speakers, "Matins" furthers the theme of embodied empathy between plants (here, a birch tree) and humans with the poet-speaker "actually curled in the split trunk, almost at peace / in the evening rain / almost able to feel / sap frothing and rising" (Glück 1992, 2, ll. 10–13). We find in this excerpt the birch voice as a polyphony; its inner physiology and outer presence in the garden space register in the "flesh" (to borrow from Merleau-Ponty) of contact between the human and vegetal. The poet concedes in another "Matins" that discourse with the birches – even of the silent, somatic kind – could be seen as a flight of fancy, but nevertheless melodramatically exhorts her critics (her husband, her son, her father, her God, etc.) to "bury me with the Romantics / their [the birches'] pointed yellow leaves / falling and covering me" (Glück 1992, 13, ll. 15–17). Understanding birch voice, for the poet, is an empathic nonverbal act in which the tree's ecological presence washes over her.

In these examples, plant-voice is immanent, rather than purely symbolic; multisensory rather than solely auditory; collective not merely singular. The corporeal register of plant-voice continues with "The White Rose," in which the flower summons the poet-gardener with its elegant gesticulations. At the same time, the rose laments its inability to conceal itself, that is, to turn its "voice/presence" off:

> I am not like you, I have only
> my body for a voice; I can't
> disappear into silence—
>
> And in the cold morning
> over the dark surface of the earth
> echoes of my voice drift,
> whiteness steadily absorbed into darkness. (Glück 1992, 47, ll. 12–18)

In contrasting its linguisticity to the poet's, the rose asserts its right to expression in the mode that is particular to it. The rose voice is modulated by its earthly habitus rather than its own volitions. "Scilla" goes so far as to chide the poet-gardener for clinging to the belief that voice involves a human subject seeking a personal God. Instead, the flower speaks passionately for the collectivity of the garden through its body language: "Not I, you idiot, not self, but we, we – waves / of sky blue like / a critique of heaven" (Glück 1992, 14, ll. 1–3). For the rose-speaker, individuated voice (*i.e.* me speaking) is unthinkable: "why / do you treasure your voice / when to be one thing / is to be next to nothing" (Glück 1992, 14, ll. 3–6). As a nexus of things, feelings, and memories, voice is ecological; it is neither a property of the flower nor of the poet-supplicant nor the garden itself, but of their interdependencies.

The posthumanist concept of plant-voice that is strongly evident in Glück's *The Wild Iris* is neither metaphorical nor contingent on human vocalities. It links speaking to bodily presencing in a place (a garden) over time (the seasons). An ecology of plant-voice is relational where ecology is "thinking how all beings are interconnected, in as deep a way as possible" (Morton 2010, 255) (chapter 1). Plant-voice is not of the birch and rose themselves (as the subjects we might wish them to be), but of the syncretism of their environmental relations (the ecological beings that they already are). This idea of plant-voice reflects Rogers' proposal for a "transhuman" model of communication that overcomes the privileging of human symbol-construction and symbol-deployment (Rogers 1998). This plant-voice also builds upon scholarship in the field of human-animal communication by Emily Plec who characterizes voice as "intentional energy" exchange between humans and other forms of life. For Plec (2013, 7), an "anthropocentric grip on the symbolic" requires a "corporeal rhetorics of scent, sound, sight, touch, proximity, position, and so much more." A plant rhetorics, beyond symbolism, is fundamentally ecological, as plants negotiate their environments through their phenomenological gestures. Let us now turn to Elisabeth Bletsoe's *Pharmacopœia* to consider further examples of plant rhetorics and their silent, heteroglossic voices.

Plant Heteroglossia in Bletsoe's *Pharmacopœia*

Elisabeth Bletsoe is a British poet, born in 1960 and raised near the town of Wimborne in the district of Dorset, South West England. Her work has been associated with ecopoetics – a movement within contemporary poetry or poetic projects that investigates ecology (Bristow 2008). Recognizing the prime importance of relationships between beings, ecopoetics questions, resists, and recasts the role of human language in the domination (or appreciation) of nature. Offering a far less idyllic image of landscape than its Romanticist antecedents, ecopoetics radicalizes the nature poetry tradition, bringing ethical and relational concerns to the fore and using the language of science to do so. Jonathan Skinner offers a four-part taxonomy of ecopoetics as *topological* (referring to a topos beyond the poem); *tropological* (hybridizing poetic and scientific languages and imparting an ecosystemic quality to a poem); *entropological* (involving or mirroring ecological materials and processes); and *ethnological* (recognizing human culture within a topos) (Williams 2011, 158). With respect for nature and in a tone of humility, ecopoetic works tend to reflect principles of ecocentrism and interdependence, while developing critiques of technology and "hyperrationality" (Bryson 2002, 5–6). Bletsoe's poetry adheres to Skinner's criteria through its focus on the landscapes of South West England, specifically its wild (and weedy) flora. In its ecocentrism, her work is also deeply human.

Originally published in 1999, *Pharmacopœia* is a sequence of eleven poems, each titled with the common and scientific names of medicinal plants, as well as their locations: for example, "Stinking Iris (*Iris foetidissima*), Kilve)" (Bletsoe 2010, 101). In addition to *Pharmacopœia*, "The Leafy Speaker" (Bletsoe 2010, 75–78) from *The Regardians* (1993) demonstrates Bletsoe's particular engagement with the voices of British flora. The works' polyphonies are at once scientific, historical, and embodied. As a poetic interpretation of traditional herbal texts, *Pharmacopœia* exhibits a "double relation" of acting (humans appropriating plants as materials) and being acted upon (plants affecting human bodies by releasing therapeutic compounds or essences) (Smith 2012, 189). Bletsoe's use of the term *pharmacopoeia*, a book giving directions for the preparation of plants, animals, and minerals for medicine, positions her work within this tradition dating back to the 1st century AD text *De Materia Medica* (Dioscorides Pedanius (of Anazarbos.) 2005). However, central to her ecopoetics is the calling into question the utilitarianism that reduces flora to medicinal substances (see chapter 2). Instead, her poetry widens "our vision of each plant through reference to its multiple names, its places and conditions and its mythologies" (Tarlo 2011, 14). Interspersed with voices of medieval herbalists, the botanical voicing in *Pharmacopœia* contrasts to Glück's tacit invocation of myths in the economic diction of *The Wild Iris*. Unlike the latter, Bletsoe's voicing refrains from speaking for plants as personae, instead engendering the poet's dialogue with historical texts and the sensuous presence of plants before her.

The plant-voice of Bletsoe's ecopoetics is felt, smelled, tasted, imbibed, eaten, touched, and seen in relation to locales within South West England. The articulation of voice through plant gesture is prominent in *Pharmacopœia* and especially in "Stinking Iris." The olfactory *effluence* of the plant (to borrow from Thoreau) is denoted by its species name *foetidissima*, from Latin for "stinking." By the poem's end, the iris has undergone a poietic transubstantiation, "growing more grateful & aromatic / as it dries" (Bletsoe 2010, 101, ll. 24–25) with the closing lines drawn from the eighteenth-century herbal *Botanicum Officinale* (Miller 1722, 13). The visual rhetorics of the iris flower are likened to a form of environmental writing – "a 'pencilling' of purple-gray / blue-gray / on tombs at Carnac" – where the gerundial "pencilling" is also derived from an historic source (Bletsoe 2010, 101, ll. 19–21). Another poem in the sequence, "Elder (*Sambucus nigra*) Culbone," concerns the rich folk knowledge surrounding the species – "by yeast / & muscatel" – in which its berries have been fermented for medicinal brews (Bletsoe 2010, 104, ll. 26–27). The tactile act of wine-making invokes the animist traditions of the elder that predate Christianity and the Linnaean taxonomizing of its corporeal language. The berry is transubstantiated to wine, bringing the human and the elder into an embodied double relation "where / vortices / of 5-petalled flowers / brush lips / skin / hair" (Bletsoe 2010, 104, ll. 21–26). The mildly intoxicating concoction becomes a medium of co-voicing between the elder and the poet-herbalist, as evident in the poem's closing line: "we are both now / *forspoken*" (Bletsoe 2010, 104, ll. 29–30). This ecology of voice resists a clear boundary between internal and external plant-voice – that is, the voice of the plant and the voice given to it, respectively. Through an ecopoetic practice, human speakers of the past and present come into unruly dialogue (broadly conceived and not solely verbal) with Bletsoe's elder of Culbone hamlet.

The theme of botanical folklore is strongly evident in "The Leafy Speaker," a wide-ranging reflection on the archangel Raphael as a healer and patron of writers, the Roman god of medicine Mercury, the Green Manplant deities, and oak-related mythologies (Bletsoe 2010, 75–78). The poem reinterprets the rich folkloric tradition of the oak. For example, nymphs of oaks were the attendants of the Greek deity Artemis. It was Erysichthon who destroyed the fertility goddess Demeter's oak grove and was therefore cursed to endure insufferable hunger. Roman emperor Theodosius the Great's anti-pagan campaign involved destroying the oak at the oracle Dodona in 391 AD (Young 2013, 106). In the poem, these mythological figures and moments coalesce in a presence of a roadside oak and "call to me: / a startling epiphany / though your feet seem solid enough / on the unremarkable pavement" (Bletsoe 2010, 75, ll. 4–7). The oak's spatial and historical articulations are likened to language in the following passage:

> branches
>   in rhyming couplets
> stream from your mouth
> leafily speaking
> the greene man the holy oake
> all that is ancient and mute
> you give tongue to:
> a vision of the world
> before the world. (Bletsoe 2010, 77, ll. 85–93)

The leafy speaking of the oak is not in an aural mode of ontology. The oak is not merely mouthing history. The "calling" happens as sensation and through the oak's expression of its physical being and mythological relations to "all that is ancient and mute." The oak-voice, like Glück's wild iris, is prediscursive and heteroglossic.

If we cannot hear plant-voice, then it *must* exist. Bletsoe's ecopoetic practice makes this clear by giving nuance to the silent voices of plants. It shifts us away from conventional notions of voice as the mechanical outcome of the vocal chords; as the result of the consciousness of well-brained organisms; and as a purely aural expression privileging those with vocal capacities. The historicized plant-voice in her work resists a mechanistic paradigm in which "the voice arises in roughly the same manner for everyone. Air moving along the chords of the voice box causes vibration like the river wind against a simple reed" (Appelbaum 1990, 3). What emerges, instead, is the location of the expressions of being – sensory, ecological, historical – that constitute vegetal situatedness. Building on Watts' notion of affect, this is voice as material enunciation. The plant-voice of Bletsoe's poetry is the vegetal announcing its being – the uttering of "the [plant] body's sensory experience of its environment and of others" (Watts 2001, 180). Rather than a barrier to voice, such poetry facilitates our attentiveness to its emergence. Her ecopoetic work decouples voice from pathetic fallacy by showing that plant-voice is less ethically fraught (*i.e.* as speaking in the place of plants) if relationally and historically grounded; and that voice is always already a heteroglossia and, as an "original impulse of being," resists its own reduction to speaking.

CONCLUSION

The voice of plants (internal to them) and the giving of voice to plants (external to them) are intrinsically related in an interplay between human and vegetal voicing. Nonverbal, ecological voice – the core principle I have been developing in this chapter in response to pathetic fallacy – is manifested through vegetal presence and human recognition of it; through taste, smell, touch, and proprioception. As Michael Marder (2013, 75) comments, "plants, like all living beings, articulate themselves spatially; in a body language free from gestures." Contrary to the second part of Marder's assertion, the poetry of Glück and Bletsoe demonstrates that the spatial articulation of plants in languages is replete with gestures. These gestures call us neither to the past nor to the future, but to the present in which all possibility inheres. They "announces the [vegetal] body's presence" (Watts 2001, 180) not by summoning through sound but by enunciating in the world's substance. As ecological presencing, plant-voice is not only the susurration of speaking subjects, bearing vocal chords and enacting discrete subjectivities. Nature speaks in the entanglement of time, space, spirit, and materiality, but not as literary conventions, communications models, and human paradigms of vocalization would have it. Rather than refusing to engage for fear of ventriloquizing them, the poetry of Glück and Bletsoe reveals the diverse expressions of plant-voice – in its silence and possibility, and as a part of our and their being.

CHAPTER 11

Tolkien's Sonic Trees and Perfumed Herbs: Plant Intelligence in Middle-earth

INTRODUCTION: TOLKIEN'S PLANTS IN ALL SENSES

Real, imaginary, and semi-fictional plants populate J.R.R. Tolkien's legendarium. Some are vocal and menacing, while others are fragrant and therapeutic. The plants (or plant-like beings) that murmur, speak, or sing most commonly appear in the form of trees. A prominent example comes from *The Fellowship of the Ring* in which the hobbit Frodo Baggins is put under a soporific spell in the Old Forest by the wrathful singing and chanting of Old Man Willow (Tolkien 1994b, 116-117). In contrast to vocal trees, plants that cannot make sounds of their own volition tend to appeal strongly through smell. For instance, one of the most celebrated plants in the legendarium is *pipe-weed* or *leaf*, based on the botanical genus *Nicotiana*. Hobbits were the first to smoke its burning leaves to relax their travel-worn bodies, heal injuries, promote clarity of mind, and foster conviviality. The inhabitants of the Middle-earth kingdom Gondor called the herb *sweet galenas* and esteemed the fragrance of its flowers but did not consume it like tobacco, as hobbits and, later, dwarves and wizards would (Tolkien 1994b, 7–8).

The examples of Old Man Willow and pipe-weed indicate the range of plant representations in Middle-earth – from "siren-like" (Brisbois 2005, 211) trees that use sonic forms of address to herbs of sociality that are pleasing to olfaction. On the one hand, these variations reflect the narrative purpose of each plant in Tolkien's fictional world. On the other, the author's knowledge of botany and ethical values infuse his rendering of Middle-earth flora and their sensory potentialities and modes of expression. Indeed, Tolkien had an ongoing interest in flora. He was an ardent gardener and especially defensive of trees (1981, 402–403, 420). The story "Leaf by Niggle" was penned as an indignant response to the cutting down of an old poplar near his home (Tolkien 1964). His attention to botany was sustained through the visual appeal of illustrated florals, the tactile experience of growing flowers and herbs, and general curiosity about the particular temporal and ontological modes of plants. Tolkien recognized "the mystery of pattern/design" in plants and described the variations between botanical families as rousing "in me visions of kinship and descent through great ages" (Tolkien 1981, 402). On occasion, he also addressed flowers directly and speculated about their capacity to respond, as an extract from his personal letters about elvish daisies discloses (Tolkien 1981, 403).

The word for plants in Tolkien's fictional language quenya is *olvar*, in contrast to *kelvar* for animals. The terms are broadly comparable to *flora* and *fauna* (Tyler 1979, 313). In *The Plants of Middle-earth* (2006), Dinah Hazell outlines the symbolic meanings, mythological associations, narrative elements, and historical allusions of Tolkien's olvar. However, I suggest that the communicative abilities of Middle-earth flora also prompt readers to think imaginatively beyond the prevailing conception of the plant as mute and unintelligent – as a relatively immobile life form defined in its other-than-animalness (Marder 2013a, 2). Some of Tolkien's olvar are hybrids between actually existing species and mythological and folkloric plant personae. While historically and ecologically grounded, his botanical legendarium, in part, also prefigures contemporary evidence concerning acoustics, consciousness, and memory in the vegetal world. In the light of this scientific research, Tolkien's olvar might not be purely imaginative or metaphorical after all, but rooted, at least partially, in emerging empirical findings. As an aspect of their behavioral ecology, plants have been shown to emit and respond to sound frequencies (Gagliano 2013, Gagliano, Mancuso, and Robert 2012).

In this chapter, recent research into plant science will be applied as a framework for investigating plant representation in Middle-earth. Two divergent examples from Tolkien's vegetal realm will be selected. The first example comprises Tolkien's sonic trees and arborescent beings, including Old Man Willow, huorns, and ents. So as to highlight differences of representation, the second focuses on the healing herb *athelas*, also known as *kingsfoil*. The former does not vocalize like the trees but effuses fragrantly and employs non-sonic, multi-sensory modes of communication. Using the scientific research as a foundation for textual analysis, the distinction between *extrinsic* and *intrinsic* plant capacities will be developed. However, ultimately Tolkien's affection for plants is not without its shortcomings. An arborescent environmental ethics privileges venerable trees (conferring to them the ability to vocalize) and bars herbaceous plants from similar modes of consciousness and memory through the intentional use of sound. Despite Tolkien's sympathy for the vegetal and his valuing of plant intelligence in his fictional universe, his tree-based ethics internalizes a chain-of-being conceptualization of the botanical world that places trees well above herbaceous plants.

VEGETAL INTELLIGENCE ON EARTH AND MIDDLE-EARTH

The ascription of intelligence to trees through their ability to speak can be traced to the prehistorical world, or Arda, of Tolkien's legendarium. Yavanna is the creation divinity responsible for growth and life in Arda. Yavanna, whose name means "Giver of Fruits" (Tolkien 1977, 27), made the first *kelvar* and *olvar* in Middle-earth. She is famous for creating The Two Trees of Valinor known as Telperion (the elder of the two whose flowers emit silver light) and Laurelin (with golden leaves) (Tyler 1979, 647). Yavanna evokes the fertility goddess figure of world mythologies, notably the Old Norse goddess Freya and the Roman Venus (Dickerson and Evans 2006, 120). Through Yavanna's prayers, all plant and animal life comes into existence, but her special affinity is for trees. In *The Silmarillion*, Tolkien describes her as:

> the lover of all things that grow in the earth, and all their countless forms
> she holds in her mind, from the trees like towers in forests long ago to
> the moss upon stones or the small and secret things in the mould.
> (Tolkien 1977, 27)

Into the barren terrain of Arda, Yavanna propagates seeds that germinate into a variety of "growing things great and small, mosses and grasses and great ferns, and trees whose tops were crowned with cloud as they were living mountains" (Tolkien 1977, 35). In addition to disseminating seeds, Yavanna is able to assume an arboreal form herself, appearing sometimes as "a tree under heaven, crowned with the Sun" (Tolkien 1977, 28). After the mighty spirit Melkor brought widespread devastation to Arda, Yavanna laments the fate of the trees to Manwë, lord of the creation beings known as Ainur (of which she is one). She implores Manwë to bestow the power of speech to trees: "Would that the trees might speak on behalf of all things that have roots, and punish those that wrong them!" (Tolkien 1977, 45). At first considering her appeal a "strange thought," Manwë is then reassured by Yavanna that the great trees already possess the capacity for divine song, having intoned their praises to Ilúvatar, the Creator, during the making of the clouds and rain (Tolkien 1977, 45–46). Yavanna's request for a guardian of the trees is granted by Manwë. He creates the arborescent ents, the "Shepherds of the Trees" including Treebeard, who protect the forest from harm and bear the gift of vocalization.

Yavanna favors the trees of Middle-earth. The arborescent Ents are brought into existence to speak on behalf of the forest and to ensure its longevity. However, some trees, notably Old Man Willow, can vocalize without the intervention of other beings, raising the possibility of intrinsic consciousness and memory in the botanical legendarium. Although devised for a fictional universe, Tolkien's singing trees have some empirical underpinnings. A body of evidence from contemporary plant science recognizes plants as greatly sensitive organisms that "perceive, assess, interact and even facilitate each other's life by actively acquiring information from their environment" (Gagliano, Mancuso, and Robert 2012, 3). In particular, *plant bioacoustics* describes their perception and production of sound, demonstrating that plants emit sonic patterns and modify their behaviors in response to received auditory stimuli (Gagliano 2013, Gagliano, Mancuso, and Robert 2012). Although "vanishingly small," plant acoustic emissions can take the form of "substrate vibrations or airborne sounds" (Gagliano, Mancuso, and Robert 2012, 1). This understanding of sound ecology counters the conventional bioacoustic model that tends to attribute the vocalizations of plants to sudden pressure changes in their water systems. Accordingly, sound in plants has been regarded as incidental. This reflects the dominance of the chemical model of plant communication, based on hormonal emissions. In contrast, recent bioacoustic research suggests that vocalization is an active process that facilitates more efficient signaling than chemical transmission in plants. As such, sound serves an ecological function with consequences for fitness and as part of the plant's evolutionary constitution (Gagliano 2013, 1).

An active model of phytoacoustics counters the prevalence of the chemical messaging paradigm and supports the broader reinterpretation of plants. Other scientists argue that plants exhibit features of intelligent behavior (Trewavas 2014), including the ability to learn and remember (Gagliano et al. 2013). The concepts of plant learning and memory were first put forward by the Bengali biologist Jagadish Chandra Bose (1858–1937), who also suggested that plants have nervous systems though they lack nerves and neurons (Trewavas 2014, 18). Part of the adaptive behavior of a plant is mnemonic, involving the ability to recall and respond to biological information (Gagliano et al. 2013, 63–64). Species such as *Mimosa pudica* even display long-term recall, in which "an enduring memory of a past event" (Gagliano et al. 2013, 69) entails responding to cues and altering behaviors to enhance survival. Trewavas states that:

> No wild plant could survive without a memory of its current perceived signals or without a cumulative memory that collates its past information experience and integrates it with present conditions so that the probabilities of potential futures could be assessed. (2014, 90)

However, traditional learning research in biology, based on neuronal processes, excludes organisms like plants from the behavioral domain of learning.

The vocalizations of some Middle-earth trees and arborescent beings bestow – at least to this segment of the vegetal realm – aspects of behavior through memory and learning. Within Tolkien's botanical mythologization inheres the freedom of imagination and thinking necessary to reconceptualize the capacities of plants and to disrupt narrow, animalistic paradigms of intelligence. However, as I will later argue, this potential is only partially realized in Tolkien's work and is, instead, constrained by an arboreal ethics. Recent scholarship in plant studies critiques the historically prevailing attitudes toward plants in the disciplines of science, philosophy, and literature (Hall 2011, Marder 2013a, Ryan 2012a). Plants have been considered the passive elements of landscapes (ontologically liminal, somewhere between rock and animal); the aesthetic features of scenery (with trees and other charismatic forms most appreciated); the inert materials of construction (or in Heidegger's terms "standing reserve"); or the biochemical constituents exploited as food, fibers, or medicines (the utilitarian, ethno-pharmaceutical model of plants as resources).

Yet, plants constitute the vast majority of the world's living things; the global terrestrial plant biomass (phytomass) might be one-thousand times greater than animal biomass (zoomass), although estimates are highly variable and measuring techniques unreliable (Smil 2008, 80). As indicated by the linkage between learning and neuronal systems, zoocentric biases toward plants can marginalize their real capacities. Informed by research into plant intelligence and behavior, critical studies of the vegetal call into question conceptualizations of plants that are founded on their presumed deficiencies. As plants are shown to manifest attributes of intelligent behavior, their ethical representation should become a more pressing concern.

Considering this background, it is productive, then, to discriminate in broad terms between the *extrinsic* and *intrinsic* capacities of plants. *Extrinsic* describes those capacities registered as environmental elements or other living beings exert force upon plants. *Intrinsic* refers to those attributes generated actively by plants in relation to their surroundings. Let us consider the value of this typology. As we have seen, bioacoustics differentiates between the reception, detection, and emission of sound by plants. On the one hand, there is receptivity to acoustic signals from the environment; on the other, there is the generation and moderation of sound by plants. The latter implies an active sonicity that influences the responses of other organisms and supports the adaptation of "the plant" (the living organism and its species) over time. Indeed, research indicates that plants have the ability to perceive vibrations and change their behaviors in response to acoustic frequencies (Gagliano, Mancuso, and Robert 2012, 2). In contrast, a notion of phytoacoustics as passive describes the vocalizations of trees as the sound of the wind through their leaves, as the thump of hail pelting their canopies, or even the sudden change in water tension within the plant. The rustling of leaves, hail strikes on canopies, or changes in internal tension are mechanical rationalizations of sound in which things (*i.e.* air, water, pressure) act upon the plant – in this sense, they are largely extrinsic factors affecting sound. However, the "vanishingly small" frequencies used by trees to negotiate their life-worlds indicate an intrinsic form of sound, in which plants exert poietic forces upon their environments. Intrinsic capacities are phytocentric (or, at least, ecocentric) in that they engage a positive conceptualization in which plants can actuate through the modes that are particular to them. In *The Fellowship of the Ring*, the tension between intrinsic and extrinsic qualities surfaces as the hobbits consider the uncanny trees of the Old Forest. Merry observes that:

> the Forest *is* queer. Everything in it is very much more alive, more aware of what is going on [...] I thought all the trees were whispering to each other, passing news and plots along in an unintelligible language; and the branches swayed and groped without any wind. (Tolkien 1994b, 108)

In this passage, consciousness is linked to the vocalizations of the trees. Their whispering is not a mechanical result – for instance, the friction of the wind over the tree. The evidence observed by Merry and Pippin is that the branches sway despite the wind's absence. Hence, the forest's awareness of the hobbits is connected to its intrinsic, though indecipherable, language – that which has its provenance within the trees themselves. The obverse of hobbit (or human) consciousness of the forest is the forest's self-awareness and the actions of hobbits (or humans) within it. Similarly, there are the hobbits' recollections of the forest – the memories other conscious beings have in relation to the trees – but also the forest's remembrance of its past abuses – signifying an inherent (*i.e.* intrinsic) capacity for vegetal memory. These themes and distinctions, extending plant science to plant representation, are crucial to understanding the example of Old Man Willow discussed in the next section.

OLD MAN WILLOW AS THE EMBODIMENT OF ACTIVE NATURE

The contrast between intrinsic and extrinsic plant faculties aligns with Michael Brisbois' distinction between passive and active nature in Tolkien's work (2005,

203). Although recognizing the limitations of this binary, Brisbois characterizes *passive* as referring to the forms of nature that are most integral to Middle-earth narratives. Not necessarily the pejorative construction of nature as "inert or stagnant," passive nature is a "less manifest force" that provides realist, moral, and symbolic anchor points (Brisbois 2005, 204). By comparison, *active* nature includes "fantastic" beings (such as trees, ents, and huorns) and elements that more directly and intensively affect the course of the narrative. Echoing my notion of intrinsic plant capacities, active nature exhibits qualities of perspective, intelligence, and sentience (204). Brisbois further divides these categories four ways. *Passive* includes "essential" nature underlying the realistic features of Middle-earth and "ambient nature" evoking an atmosphere of morality, divinity, or spirituality (204). Additionally, *active* comprises "independent" nature that is intelligent but reclusive, living apart from culture, while "wrathful" nature is marked by aggressiveness and malicious interaction with Middle-earth inhabitants (208). In less typological terms but nevertheless raising the passive/active distinction, Dinah Hazell comments that:

> not all trees in Middle-earth are animated. Frequently they are unidentified and form part of the landscape and occasionally serve functional purposes as landmarks, vantage points, protection from weather and visibility, sleeping quarters, safety from attack, and even battering rams as used by the Orcs at Helm's Deep. (Hazell 2006, 68)

Hazell implies that while some *olvar* exhibit intelligence, consciousness, sentience, and memory, as expressions of active nature with intrinsic capacities, others are exclusively acted upon by Middle-earth's denizens as essential or ambient nature. Hence, there are degrees of plant liveliness in Tolkien's legendarium, from common diminutive herbs without vocalizations to lofty trees communicating their presence and states of mind (Walker 2010, 45). In *The Fellowship of the Ring*, as the company left the safety of the garden Lórien, "in the trees above them many voices were murmuring and singing"(Tolkien 1994b, 361). Moreover, in *The Hobbit*, Bilbo Baggins and companions observe the percipience of Mirkwood Forest as "the trees leaned over them and listened" (Tolkien 1995, 130).

These differences – between active/passive nature, intrinsic/extrinsic capacities – are especially evident in Old Man Willow and his spiteful singing. The Old Forest of Eriador is a vestige of an immense primeval tract of the Elder Days (Tolkien 1994b, 107–120). While most of the vegetation – including quintessentially European and North American grasses, nettles, thistles, hemlocks, pines, oaks, and ashes – is "passive essential" nature, other trees are mobile, conscious, and sentient. As intelligent beings with intrinsic, intelligent faculties, the willow trees in particular hold memories of having been overrun and destroyed, and thus harbor malice toward freely moving beings, such as the hobbits (Hazell 2006, 73). We are told that long ago, "the hobbits came and cut down hundreds of trees, and made a great bonfire in the Forest (Tolkien 1994b, 108). Old Man Willow, who lives on the banks of the Withywindle ("Winding-willow") River, is the most commanding of the trees – a fantastic being – and embodies the malevolence of the forest or, what Brisbois (2005, 208) calls, the "active wrathful" nature of Middle-earth. He is the first antagonist encountered by the hobbits as they venture out of the Shire, past the protection of the Hedge, and into the forest. As the hobbits enter the Withywindle Valley, they perceive the comforting *extrinsic* sounds of the forest, as the wind acts upon the vegetation: "the reeds were rustling, and the willow-boughs were creaking" (Tolkien 1994b, 113). However, their moment of respite in the dreary and menacing forest is interrupted by Old Man Willow's sleep-inducing lull:

> Suddenly Frodo himself felt sleep overwhelming him. His head swam. There now seemed hardly a sound in the air. The flies had stopped buzzing. Only a gentle noise on the edge of hearing, a soft fluttering as of a song half whispered, seemed to stir in the boughs above. He lifted his heavy eyes and saw leaning over him a huge willow-tree, old and hoary. (Tolkien 1994b, 114)

Here, Frodo's perception shifts from the wind as an acoustic agent to sound as a sensation that registers in his body as an opiate does. The narrative tension in the Old Forest is indeed one of sonicity, in which extrinsic sound (the creaking boughs) transforms unexpectedly into intrinsic vocalizations (the soporific Siren song of the willow). The latter signify the sentience of the forest. When Frodo kicks a tree in protest of their malevolence, "a hardly perceptible shiver ran through the stem and up into the branches" accompanied by its faint laughter (Tolkien 1994b, 115). After attempting to induce sleep in the hobbits, the willow tries to strangle Merry and Pippin, overpowering their ability to speak with its increasingly clamorous and agitated voice.

However, the intervention of Tom Bombadil, the eldest of all creatures in Middle-earth, saves the hobbits and thwarts the willows with song: "I'll sing his roots off. I'll sing a wind up and blow leaf and branch away. Old Man Willow!" (Tolkien 1994b, 117). Tom Bombadil admonishes Old Man Willow but suggests that, although his animosity is justified, his actions against the hobbits are reprehensible (Brisbois 2005, 209). The Great Willow threatens mobile creatures because the memory of the forest's abuse. The active wrathful nature of the trees is, therefore, linked to habitat destruction. Tom's elaboration of the environmental history of the forest seems to engender empathy among the hobbits for the willow's plight:

> As they listened, they began to understand the lives of the Forest, apart from themselves, indeed to feel themselves as the strangers where all other things were at home [...] Tom's words laid bare the hearts of trees and their thoughts, which were often dark and strange, and filled with a hatred of things that go free upon the earth, gnawing, biting, breaking, hacking, burning: destroyers and usurpers. It was not called the Old Forest without reason, for it was indeed ancient, a survivor of vast forgotten woods; and in it there lived yet, ageing no quicker than the hills, the fathers of the fathers of trees, remembering times when they were lords. (Tolkien 1994b, 127–128)

Indeed, trees embody ecological ethics in Tolkien's work (Brisbois 2005, 200). In a letter to the *Daily Telegraph* in 1972, Tolkien states that "I take the part of trees as against all their enemies [...] the Old Forest was hostile to two legged creatures because of the memory of many injuries" (Tolkien 1981, 419). He also explains that the trees in his legendarium are "awakening to consciousness of themselves" (Tolkien 1981, 419) – and an integral aspect of their awakening is their intrinsic ability to vocalize.

The fathers of the trees are the ents – the conscious arborescent beings created by Manwe, at Yavanna's request, to shield the forest from further injury. Old Man Willow is closely related to the ents and huorns, especially through their shared capacities of aurality (speaking, singing, laughing, shuddering) and locomotion. Like ents, huorns resemble trees and have the power of speech and mobility. As either sentient trees or regressed ents (their exact ontology is not clear), huorns can move quickly from place to place under the shadows of true trees, but are always under the control of ents (Tyler 1979, 295). In the language *quenya*, the huorns of Fanghorn Forest, located on the eastern side of the Misty Mountains, were previously known as *galbedirs*, *ornomi*, and *lamorni*, all of which mean "talking trees." The name *lamorni* is derived from *lam* for "sound" and *orni* for "trees." Thus, the huorns are literally "trees with voices" (Tolkien 1990, 55). According to the ent Treebeard, the chief guardian of Fanghorn Forest, the huorns, as Merry relates, "still have voices, and can speak with the Ents [...] but they have become queer and wild" (Tolkien 1994c, 551). Ents "have become almost like trees, at least to look at. They stand here and there in the wood or under its eaves, silent, watching endlessly over the trees" (Tolkien 1994c, 551). While other plants in Middle-earth possess supernatural or healing abilities, it is this tribe of beings – ents and huorns, including Old Man Willow – who speak in this way. But their speaking tends to come as an inversion of expectation to Middle-earth inhabitants, as "the rustling of the wind" turns into the horror of "great groping trees all around you" (Tolkien 1994c, 551); as extrinsic becomes intrinsic sound. This reflects an assertion by ecocritic Timothy Clark (2011, 53) that "the natural world is full of indicators, signs and communications, associated with diverse and (to us) mostly opaque modes of intentionality and reference." The opacity of groping and speaking trees, even to the hobbits and others, is a significant narrative dimension of the legendarium.

Hazell (2006, 74) asserts that Old Man Willow's vocalizations reflect the behavior of actual willows as their leaves and limbs "sigh' with the wind. However, such a statement seems to gloss over Tolkien's intention to endow trees with consciousness so they might resist their exploitation by two-legged creatures. It also neglects the rich mythological traditions from which Tolkien might have drawn in devising the active Middle-earth plant legendarium. Alongside his use of sonorous willows as imaginative elements and ethical flashpoints, Tolkien was fully immersed in a literary body in which vocal trees are not unusual. It is well-known that Tolkien was inspired by early Germanic, especially Old English (mid-5th century to the Norman conquest of England in 1066), literature, poetry, and mythology. Documented influences include *Beowulf* (700–1000 AD), Norse sagas, such as the *Völsunga* and *Hervarar*, and the Middle High German epic poem *Nibelungenlied* (all three dating to the 13th century) (Carpenter 1977, 138-139, 144–145, 202). Tolkien also acknowledged the influence of non-Germanic sources, such as Sophocles' play *Oedipus the King* (first performed in 429 BC) and the nineteenth-century Finnish epic *The Kalevala*. For example, the character Väinämöinen from the latter was the model for the wizard Gandalf of *The Lord of the Rings* (Snodgrass 2010, 161–162). Additionally, in Myth of the First Trees, Väinämöinen directs Sampsa Pellervoinen to sow seeds on a barren landscape. After struggling to germinate, the oaks eventually proliferate across the country, preparing the ground for herbs, flowers, berries, and agricultural crops (Lönnrot 2007, 11). Here, oaks lack voices, but they are integral to creation.

Tolkien scholars have suggested that trees are the most potent symbols in Middle-earth (Dickerson 2013, 73–74). Therefore it is crucial to consider the possible mythological underpinnings of Middle-earth's speaking trees, bearing in mind extrinsic and intrinsic acoustics. For example, in *Metamorphoses* (8 AD) by the Roman poet Ovid, based on Virgil's *Georgics* (29 BC), the musician Orpheus laments the loss of Eurydice to the sting of an adder and strums his lyre in the abandonment of his grief. Sitting atop a verdant hill, the downhearted notes of the instrument give provenance to a forest of oaks, poplars, beech, hazel, laurels, ash, fir, maples, willows, myrtles, and elms (Ovid 2010). The story of Orpheus associates music to prolific arboreal growth, but does not indicate if the trees themselves vocalized. However, at the Greek sanctuary of Dodona at Epirus, an oracular oak was the medium for revering Zeus. In the rustling of the leaves on a still windless day, Zeus announced himself. The Sacred Oak, an ilex oak (*Quercus ilex*), is thought to have lived until 180 AD with a priestess regularly consecrating Dodona until about the third century. The crepitation of the leaves without the wind's external manipulation implies intrinsic movement and vocalization connected to its supernatural agency as the voice of Zeus (Varner 2006, 22). There are also numerous folk tales, apart from these Greco-Roman examples, of vocal spirits in forests. For example, the Scandinavian sprite Huldra lives in a pine grove and sings a lament called *huldrslaat* in a "clear and sweet" voice (Goddard 1871, 180). The pine forest is the medium for sound, rather than the means of its production.

I suggest that such currents of mythology most likely figured into Tolkien's representations of singing trees and, specifically, Old Man Willow. Of course, a segment of this tree lore – to which Tolkien could have been exposed – is specific to willows. For instance, in the second century AD, the Greek geographer Pausanias, in his book *Description of Greece*, discussed a painting of Orpheus grasping a lute in one hand and a willow branch in the other, again relating willows to music (Pausanias 2012, 545). Willows lend their name to the River Helicon, described by Pausanias as disappearing underground then rising again before descending to the sea. The women who killed Orpheus attempted to wash their hands in the river, but it disappeared beneath the ground to prevent them from doing so (Pausanias 2012, 481). In Greek mythology, Itonus or "Willow Man," son of Amphictyon and husband of the nymph Melanippe, established a sanctuary (or willow cult) of Athena, the goddess of wisdom, courage, and law (Graves 1955, 47). Furthermore, Circe, the Greek goddess of magic and daughter of Helios, kept a willow grove where the tree was associated with the moon and known for its healing properties. A more contemporaneous source for Tolkien was the work of English book illustrator Arthur Rackham, whose mythologized drawings of trees, including those in Kenneth Grahame's *The Wind in the*

Willows, influenced the depiction of Old Man Willow as active nature (Carpenter 1977, 162).

ATHELAS AS NON-SONIC HEALING PLANT

As a "fantastic" plant being with the powers of speech and locomotion, Old Man Willow is quintessentially active nature, bearing attributes of percipience and sentience. Sound is the dominant sense around which his intelligence is constructed. To a great extent, the uncanniness of Old Man Willow relates to his being heard before being seen, as the hobbits move through The Old Forest. Not merely "the wind in the willows," to echo Grahame and Rackham above, his faculty of vocalization involves intention (malice) and decision-making (attempting to drown, anaesthetize, and garrote them) (Flieger 2000). Turning from sound as a principle of intelligence in Middle-earth trees, this section will examine Tolkien's use of other sensorialities in representing the communicative modes of plants. Athelas, or kingsfoil, is one of the most powerful healing herbs in Tolkien's legendarium, along with the tobacco-like pipe-weed, mentioned in the introduction to this chapter (Hazell 2006, 31). Whereas Old Man Willow expresses wrathful nature, kingsfoil heals the inhabitants of Middle-earth of a variety of afflictions. Rather than passive, in the essential or ambient sense (Brisbois 2005, 204), kingsfoil's intrinsic capacities derive from taste and smell, not from an audiocentric logos. As a medicinal plant that lacks vocalization, kingsfoil epitomizes Tolkien's evocation of non-aural and non-visual senses (described in the next section as "autocentric" senses) in the plant legendarium. Tolkien privileges the sonicity of Old Man Willow and denies vocalization to non-arboreal plants.

In Tolkien's legendarium, kingsfoil is a plant of remarkable therapeutic virtue. The scent of its long leaves acts as an analgesic, antihemorrhagic, and diaphoretic, relieving pain, stemming bleeding, and warming the body, respectively. *Kingsfoil* (meaning "Kings Leaf," from the Old French word *foil* for "leaf") is the name given by the inhabitants of the kingdom Gondor to the herb athelas (Tyler 1979, 315). The name *athelas* most likely derives from the Middle English word *athel*, meaning noble by birth or character (related to the Anglosaxon term *æoele* for "noble") and *atheling* for "prince" (Hazell 2006, 32–33). In the First Age, Huan the Hound of Valinor (the greatest dog of the epoch) gifted the beautiful maiden Lúthien with a plant to heal the chieftain Beren of an arrow wound (Tolkien 1985, 269). The Númenoreans (the denizens of Westernesse) brought the herb to Middle-earth during a three-thousand year period known to the Elves and Dúnedain as *Second Age*. Connected to the

"People-of-the-West," kingsfoil is found in the North only where Númenoreans have walked, paused, or camped. The Númenoreans adopted the Valinorean botanical name *asëa aranion* (meaning "leaf of the kings") for athelas. Known in the southern kingdom Gondor as *kingsfoil*, the species grew prolifically. Its healing attributes were unknown or forgotten there, but its leaves were valued for their fragrance (Tyler 1979, 42). After Frodo is wounded by the menacing Nazgûl (or Ringwraith in the Black Speech of Mordor) on the hill Weathertop, the healer and chieftain Aragorn (also known as Strider) treats him with athelas:

> He crushed a leaf in his fingers, and it gave out a sweet and pungent fragrance. 'It is fortunate that I could find it, for it is a healing plant that the Men of the West brought to Middle-earth [...] it has great virtues, but over such a wound as this its healing powers may be small'. (Tolkien 1994b, 192)

Aragorn bathed Frodo's shoulder in a boiled infusion of athelas. The herb allayed the pain and relieved the sensation of coldness overcoming Frodo's body after the injury, but the party still had to seek out the master Elrond to heal the wound completely. Meanwhile, despite the herb's insufficiency as a cure in this instance, its refreshing fragrance soothed and cleansed the minds of the anxious party.

The chief healing characteristic of athelas is its pleasant aroma. The most detailed description of the herb comes in "The Houses of Healing" from *The Return of the King* (Tolkien 1994a). The healers of the southern kingdom of Gondor cared for the injured and sick at the Houses of Healing in Minas Tirith. Aragorn requests athelas in treating Lady Éowyn, Lord Faramir, and Merry after the Battle of the Pelennor Fields in the Third Age. Ioreth, the elderly nurse of Minas Tirith, responds that she "never heard that it had any great virtue" though its smell is sweet and wholesome (Tolkien 1994a, 846). After the herb master of the House declares he knows nothing of the weed other than its use in sweetening the air and as a folkloric remedy for headaches, Tolkien attributes the loss of traditional herbal wisdom to its characterization as old wives tales and its reduction to quaint rhymes:

> When the black breath blows
> and death's shadow grows
> and all lights pass,
> comes athelas! come athelas!
> Life to the dying
> In the king's hand lying! (Tolkien 1994a, 847)

Eventually presented with six dried leaves, Aragorn crushes the herb, filling the room with a fragrance "like a memory of dewy mornings of unshadowed sun" (Tolkien 1994a, 847). Aragorn applies the steam of two boiled athelas leaves to Éowyn's brow, restoring warmth and strength to her body. He also treats Merry, infusing the room with the fragrance of athelas, "the scent of orchards, and of heather in the sunshine full of bees" (Tolkien 1994a, 851). The healing qualities of athelas reflect Tolkien's conception of good, whereas the trees of The Old Forest are personifications of evil (Crabbe 2000, 159–160). While sound is associated with the malevolent trees – their agony and malice against their destroyers; their active wrath – smell is linked to healing and goodness. Moreover, the recollections of athelas spoken by the herb master, nurse, and others imply its mislaid healing tradition – in which the herb is an object of use and memory – but, with defiance, Old Man Willow actively *remembers* the abuses committed against the forest.

Rather than constructing the communicative capacities of the vegetal world wholly in sonic terms, it is possible to conceive of the diverse chemical vocabularies of plants (Chamovitz 2012). Researchers have described an array of volatile chemicals that enable communication within a plant and between a plant and other organisms (Karban 2016). Several thousand volatile compounds serve a number of roles, including governing interactions with herbivores, pathogens, and pollinators (Raguso and Kessler 2016). Organic cocktails of terpenes, benzenes, esters, or amines, used for antagonistic or mutualistic purposes, create the fragrance of a flower or crushed leaf (Séquin 2012, 81). Hence, while sound marks Old Man Willow's consciousness, smell signifies the intelligence of athelas; both constitute the *voicing* of Tolkien's botanical legendarium. Whereas the willow vocalizes (speaks, sings, whispers, mutters) in terms that mirror the human, athelas speaks (inverted commas not required) through an olfactory lexicon – a vocabulary of direct sensory affect related to its healing capacity and evocation of memory. As research suggests, odor-evoked memories, those triggered by olfactory experiences, are older and stronger than memories elicited through verbal and visual cues (Willander and Larsson 2006). It is mistaken, then, to conceptualize athelas and other fragrant plants as ambient passive nature, in Brisbois' terms. For instance, the thickets of *aeglos* on the lower slopes of the towering hill Amon Rûdh are described as "long-legged [creating] a gloom filled with a sweet scent" (Tolkien 1980, 99). But we have seen that, as an intrinsic faculty, the fragrance of a plant is not merely atmospheric (here, gloomy), but supportive of a plant's diverse adaptive responses to the environment.

The fragrance of athelas is sweet, invigorating, and cleansing. Its characteristics evoke those of an aromatic herbaceous perennial with which Tolkien surely would have been familiar: yarrow (*Achillea millefolium*). Known folklorically as *herbal militaris*, soldier's woundwort, nosebleed plant, milfoil, and thousand leaf, yarrow is a common wildflower and medicinal species of the English countryside. The yarrow genus, *Achillea*, derives from Achilles, the Greek hero of the Trojan War who is said to have used yarrow to heal the wounds of his soldiers (Watts 2007, 440). When Paris shot Achilles in the heel with an arrow, no treatment, including yarrow, was adequate enough to save him. Here, there are striking parallels between the tale of Achilles and Tolkien's maiden Lúthien treating the arrow wound of chieftain Beren with athelas during the First Age. When Curufin attempted to shoot Lúthien with an arrow, Beren intervened and was struck in the chest. The hound Huan brought Lúthien an herb from the forest, which "staunched Beren's wound" (Tolkien 1977, 178). Unlike Achilles who perished, Beren survived through the power of athelas as well as Lúthien's "arts and by her love" (178). Like athelas, yarrow is a styptic that halts bleeding. In *The Herball or Generall Historie of Plantes*, the sixteenth-century English herbalist John Gerard observed that "the leaves of yarrow doth close up wounds, and keepeth them from inflammation, or fiery feeling: it stancheth blood in any part of the body, and it is likewise put into bathes for women to fit in" (Gerard 1597, 915). The warming qualities of athelas – evident when Aragorn bathed Frodo's shoulder after the wound inflicted by Nazgûl – are comparable to yarrow's diaphoretic (sweat-inducing) properties that are effective in treating the flu and other febrile conditions. In addition to its medical uses, yarrow has been favored for its capacity to disperse evil spirits and divine the future in many systems of folklore around the world (Watts 2007, 440).

BEYOND AN ARBORESCENT ETHICS: LISTENING WITHOUT HEARING

The preceding discussion implies that these kinds of folkloric knowledge traditions – in which herbaceous plants express their being-in-the-world without vocalization – are tacit in Tolkien's legendarium. However, represented as passive nature with extrinsic capacities, plants such as athelas (and aeglos) are *acted upon* (gathered, processed, applied) as healing agents or medicinal substances. The privileging of sound (i.e. the power of vocalization) attributed to the willow eclipses the "opaque modes of intentionality and reference" (Clark 2011, 53), through which the non-arboreal plant world articulates its sensoriality, intelligence, consciousness, and memory. The divergence between the willow as bearer of percipience and athelas as a source (of healing, calmness, pleasantness, compounds, substances) reflects broader philosophical and historical divergences in the valuation of the senses. For example, Kant argues that smell is the lowest sense, associated with stench, filth, decay, and the disintegration of human will. A smell (nauseating or fragrant) penetrates the body largely against conscious human directive; it registers without the intervention of perception as thinking (Kant 1978, 43). The developmental psychologist Ernest Schachtel (1984, 85) adopts a Kantian view, distinguishing between the allocentric (vision and hearing) and the autocentric (gustatory, olfactory, thermal, and proprioceptive) senses. For Schachtel, as for Kant, smell is lower, primitive, and "objectless." Whereas the allocentric senses are intellectual and spiritual, the autocentric senses lack the power for objectification and are physically implicated (Schachtel 1984, 89). This dichotomy manifests in Tolkien's representation of Old Man Willow (a proper noun, a plant personae) and athelas (in the lower case; an undifferentiated, unthinking but fragrant mass).

Besides privileging the allocentric, Tolkien's ethics of nature internalize an arborescent ethics (for example, Brisbois 2005, Flieger 2000). Deleuze and Guattari (2004, 234) call into question the hierarchical basis of arborescence as "the structure or network gridding the possible." For these thinkers, the tree expresses "rigid segmentarity" and "is the knot of arborescence or principle of dichotomy" (Deleuze and Guattari 2004, 234). Arborescence encodes Enlightenment ideals of linear progress, pronounced in the taxonomic paradigm of natural science. Of course, we still love trees regardless of their arborescence, but, in contrast, the rhizome enacts horizontality and a multitude of possible unordered interconnections without hierarchies: "the rhizome, the opposite of arborescence; breaks away from arborescence" (Deleuze and Guattari 2004, 324). Aside for aspens and a few other tree species, rhizomes are principally associated with the world of herbaceous plants, such as the medicinal herbs ginseng, goldenseal, *and* yarrow. Tolkien's assignation of intelligent qualities to trees mirrors the structure of the tree itself, "gridding the possible," whereas the intelligence of healing plants is muddled in the emphasis on their utilitarian value. Why shouldn't herbs like athelas have the ability to assert their consciousness as active narrative figures in the legendarium, without necessarily communicating in vocal terms? Why shouldn't they too speak of the mistreatments that historically beset them and, in doing so, manifest their capacities for memory and affect, as science has now begun to confirm and folklore has already insinuated?

Alongside the mythological currents informing his legendarium, Tolkien's plant chain-of-being is an outcome of his fellow feeling for the iconic trees of the pastoral English landscape (Rackham 1996). In an acerbic letter to the *Daily*

Telegraph (previously quoted in this chapter), Tolkien laments "the destruction, torture and murder of trees perpetrated by private individuals and minor official bodies. The savage sound of the electric saw is never silent wherever trees are still found growing" (Tolkien 1981, 420). Perhaps to defy the savage sound of the saw Tolkien devised the savage sounds of the willow in The Old Forest. But what of the torture and murder of wetlands plants, ferns, grasses and other less stately, more diminutive herbaceous members of the forest and fields? On this topic, Tolkien is less outspoken (indeed he is silent). Perhaps this is because yarrow and other common herbal plants are weeds – those species that "obstruct our plans, and our tidy maps of the world" and inhabit "shabby surroundings" (Mabey 2010, 1, 3). They co-exist with us, rather than in forested enclaves outside of "culture." While Tolkien's empathy for trees is laudable – and surely a reason why his works are important to us today in the Anthropocene – we must bear in mind the other green things with which we are in constant symbiotic relationship. They have their own modes of expression and manners of articulation, apart from vocalization, if only we could be attentive enough to listen without hearing.

CHAPTER 12

Sense of Place and Sense of Taste: Thoreau's Botanical Aesthetics

INTRODUCTION: AESTHETICS OF PLANTS

From European to North American contexts, this chapter examines the influence of nineteenth-century naturalist Henry David Thoreau's body of writings on contemporary American environmentalism has been extensively documented and theorized by literary scholars (for example, Buell 1995, Phillips 2006). Thoreau's prose evokes the natural world in scientifically precise terms and in combination with philosophical ruminations, historical references, and aesthetic judgements (Ryan 2013, 13–25). As a transdisciplinarian, Thoreau's fascination for the local environment of Concord was not only scientific, but also cultural, historical, and spiritual. Bradley Dean (2000a, xii) sees Thoreau as a "protoecologist" whose later work anticipates the birth of modern ecology through its meticulous description of natural occurrences. Four years after Thoreau's death in 1862 from tuberculosis, the German biologist and follower of Darwin, Ernst Haeckel, would propose the neologism *Oecologie* as "the science of the relations of living organisms to the external world, their habitat, customs, energies, parasites, etc." (quoted in Worster 1994, 192). Both terms *economy* and *ecology* share the Greek root *oikos*, originally denoting the daily operations and maintenance of a family household (Worster 1994, 192). As many contemporary environmental writers have underscored, ecology is the study of the earth "household" (for example, see Snyder 1969). At the heart of Thoreau's protoecological writings is an aesthetics of the natural world and a model of posthumanist practice. His ecological aesthetics resists paradigms of beauty that privilege art over nature, humanity over nonhuman life, and vision over the non-ocular senses of sound, taste, touch, smell, and spatial orientation. Moreover, Thoreau's aesthetic approach to ecology and the natural world is an embodied – rather than visually distanced – one (Dillard 2013).

Thoreau's aesthetic engagement with nature is acutely evident in his posthumously published botanical writings composed approximately from 1859 until his death in 1862 (Dean 2000b, 274). These works include *The Dispersion of Seeds* (published as *Faith in a Seed* in 1993) and *Wild Fruits* (2000), as well as a number of essays, such as "Wild Apples," culled by Thoreau from his manuscripts and submitted to *The Atlantic Monthly* and other journals. The unfinished manuscript *Wild Fruits* principally reflects his unfulfilled desire to write a comprehensive environmental history of Concord, Massachusetts, focused on the seasonal patterns of local plants, animals, and weather. The natural phenomena observed and reflected upon in *Wild Fruits* are presented in order of their appearance from the emergence of the first fruits of spring (the winged seeds of elm trees in early May) to the last fruits of winter (the berries of *Juniper repens* in March). Thoreau referred to this ambitious undertaking as his "Kalendar," reflecting his familiarity with English gardener and diarist John Evelyn's *Kalendarium Hortense*, or *The Gardener's Almanac* (originally published in 1664), "directing what he [the gardener] is to do monthly throughout the year and what fruits and flowers are in prime" (Evelyn 1699, frontispiece). As evident in these two manuscripts, Thoreau's botanical writings consist of factual information about the size, shape, and distribution of fruits blended with subjective, embodied, aesthetic, historical, environmental, and even political observations. The blended prose of *Wild Fruits* includes the visual accounting of botanical characteristics, reflections on historical sources such as the works of the sixteenth-century botanist and herbalist John Gerard, and evocations of nibbling, tasting, or consuming berries *in toto*. More importantly, plants in his oeuvre are not treated merely as the objects of scientific evaluation or visual appeal, but as subjects of complex embodied and multi-sensory human exploration of the natural world (Ryan 2012, chapter 1).

This final chapter of *Posthuman Plants* will examine Thoreau's aesthetics of gustation – of taste – in *Wild Fruits* and, more specifically, his use of poetic language to express aesthetic experiences of tasting, sampling, eating, or rejecting as unpalatable Concord's local fruits. Thoreau's ecological gustation intersects with French philosopher Michel Serres' claim in *The Five Senses* (2008) that language mediates the sensory world and brings aesthetic experiences to reflective consciousness. Serres (2008, 153) rejects the historical understanding of taste as a base or primitive sense, "the least aesthetic" of the five. The problem of taste, as such, is one of language. "Taste is rarely conveyed well [as though] language allowed it no voice," Serres claims, because "the mouth of discourse excludes the mouth of taste, expels it from discourse" (2008, 153). Thoreau's gustatory writings in *Wild Fruits* return discourse to taste and give voice to experiences of consuming the botanical environment – perhaps the most sensuous and aesthetically rich interaction one can have with nature. In doing so, his writings affirm the environment as a valid subject of aesthetic inquiry but also the sense of taste as an appropriate faculty of appreciation. Furthermore, rather than base or unrefined, Thoreau's ecological gustation, as presented in the text, is nuanced and discerning. Through taste, Thoreau distinguished the relative virtues of wild fruits, for example, considering the "bitter-sweet of a white acorn" more pleasing than "a slice of imported pine-apple" (2000, 3). We thus find proto-bioregional traces in *Wild Fruits*, praising the consumption of local foods and constructing *sense of place* through *sense of taste*, in this context, achieved through indulging in the pleasures of uncultivated fruits. Indeed, Thoreau's critique of the aesthetic sensibilities of his era, particularly "the coarse palates [that] fail to perceive" (2000, 75) the flavors of wild fruits could be relevant to us today.

SCIENCE, SENSE, AND SEXUALITY: AN EMBODIED AESTHETICS OF FLORA

For Thoreau, the beauty of nature always inherently exceeds that of art. As apparent in his "scattered remarks on problems of aesthetics" (Horstmann quoted in Bock 2008, 100), Thoreau maintained a critical posture both toward humanist aesthetics that place art above nature and, later in his work, toward the picturesque preoccupation with vistas. Landscape art, such as that of the Italian painters Guido Reni and Titian, whom he mentions in his journal, should not be conflated with nature as "bald imitation or rival" (Thoreau 2005, 176). Reflecting the development of aesthetic sensibilities linked to place consciousness, Thoreau largely dismissed English art critic John Ruskin's *Modern Painters* (originally published in 1843) as a book that "does not describe Nature as Nature, but as Turner painted her, and though the work betrays that he has given a close attention to Nature, it appears to have been with an artist's and critic's design" (Thoreau 2009, 458). Rather than nature represented or mediated by artists and in works of art, Thoreau became acutely interested in immanent nature "as she is" (2009, 458) – an aesthetics of direct contact with the world that would come to underpin the detailed botanical writings of *Wild Fruits*. It is also documented that Thoreau studied the works of English artist and theorist William Gilpin, credited with conceiving of the idea of the picturesque, "at greater length [...] than any other non-contemporary figure" (Harding and Meyer quoted in Conron 2000, 291). Through the positive effect of Gilpin's insistence on language as a fitting medium for picturesque representation (analogous to paint itself, so Gilpin suggested), Thoreau's journal underwent a transformation from the mundane accounting of facts and occurrences to the intricate textual illustration of natural phenomena and their cultural contexts. Additionally, Thoreau was also known to carry a Claude glass (a small convex mirror that imparts a painterly ambience to what is viewed) as part of his recording of the Concord environment (human and nonhuman, natural and cultural) in his notebook (Conron 2000, 292–294). However, unlike Gilpin's emphasis on scenic grandeur and broad vistas, Thoreau would eventually gravitate toward less picturesque landscapes, such as wetlands, with keen attention to recording their minutiae and temporal changes. Hence, it could be said that one reason for Thoreau's departure from the picturesque aesthetic was geographic. The landscape of Concord is sufficiently different to that of Gilpin's English countryside and necessitates aesthetic ideas and approaches that deviate from conventions formulated elsewhere. Coming to regard the vista as an outmoded Romantic preoccupation, Thoreau attended to the minuscule detail of his environs, developing an attentive practice of multi-sensory environmental portraiture that reaches its zenith in *Wild Fruits*.

Thoreau's reconfiguration of humanist aesthetics and his valuing of wild nature over art were critical to his development of a particularly American mode of environmental thought and representation in the late nineteenth and early twentieth centuries (Davis 2010, 563). Specifically, his botanical aesthetics involve visual appreciation of flora in close connection to sensorial interaction with plants and their environments (Berleant n.d.). His mode of immersive corporeal engagement with the botanical world resists the predominantly ocular approach of scientific authority, as achieved in taxonomic classification and morphological description. Thoreau's embodied aesthetics of plants also intensely contrast to Kantian formalism and, principally, the contested notion of "disinterestedness." Based in the paradigm of aesthetics as a "science" of sensory perception, this principle dictates that "the pleasure which grounds a judgment of taste should not be desire-related" (Janaway 2003, 71) and, even in "strong, moderate, and weak" forms, seeks to exclude highly subjective or idiosyncratic reactions to art and nature (Kreitman 2006). Abandoning the possibility of disinterestedness and the detached aesthetics of the picturesque, Thoreau affirms that our aesthetic tastes originate in our bodies in vibrant relation to nature (Phillips 2006, 303). Whereas Kant devalues sensuous experiences of human pleasure, particularly eating, Thoreau embraces them as part of a corporeal epistemology of the environment – one particularly centering on knowledge gained through acts of tasting, smelling, and touching.

Indeed, rejecting Kantian skepticism, Thoreau (1993, 26) adopted a form of sensuous and even erotic empiricism involving contact with nature through "the bodily eye." Rather than treating imagination and understanding, the body and the mind, science and art, as opposed terms, Thoreau sought their complementariness (Phillips 2006, 302). His perceptions of the environment are direct, affective, and, at times, idiosyncratic – in other words, anti-Kantian in their subjectivity. He recognizes the immanence of nature and resists its reduction to the moralistic symbols or figures of transcendence that define the Romanticist version of nature (Phillips 2006, 304). The sensuous "aesthetics of engagement" evident in his work regard the natural world as an active phenomenon – one that is contingent on human interactions with other living beings, natural elements, and ecological processes (Berleant 1991). Dana Phillips (2006, 301) argues that the aesthetic and the erotic intermingle in Thoreau's prose, resulting in "an aesthetics of sheer sensual abandon." For example, regarding high blueberries (*Vaccinium corymbosum*), the bushes during winter bend over "nearly to the ice [...] with lusty young shoots running up perpendicularly by their sides, like erect men destined to perpetuate the family by the side of their stooping sires" (Thoreau 2000, 34). In addition to imparting humor and lightness to the text, the eroticizing of plants reflects Thoreau's "embattled approach" to scientific knowledge, with which he was both conversant and critical (Phillips 2006, 299). Indeed, alongside his use of caricaturization and eroticization, he consistently inflects scientific understandings of plants in *Wild Fruits*, even speculating on the exact taxonomy of certain species, including a variety of high blueberry: "narrow leaves, and a conspicuous calyx, which appears to be intermediate between this and the *Vaccinium vacillans* or *Vaccinium pennsylvanicum*" (Thoreau 2000, 34). His prose (in its more descriptive and perhaps mundane moments) demonstrates an awareness of the botanical knowledge of his era, especially the taxonomic relationships between plants: "Huckleberries are classed by botanists with the cranberries (both bog and mountain) [...] plants of this order (*Ericaceae*) are said to be among the earliest ones found in a fossil state" (40).

While Thoreau's visual perception was acute, as exemplified in his careful observations of high blueberry, his prose shuns distanced ocular representation for more intimate contact with the environment. As some critics have observed, an aesthetics of the natural ornament can be found in his writing (Davis 2010). For example, Thoreau describes the berries of red osier dogwood (*Cornus sericea*) as "the pendant jewellery of the season dangling over the face of the river and reflected in it" (2000, 123). Represented as an ornament, nature reflects the balance, symmetry, and pleasing coloration of aesthetic beauty, or, conversely, nature becomes a template or model for the human creation of non-living ornaments (Davis 2010, 561). However, rather than internalizing a concept of stasis, Thoreau's aesthetics of the ornament involve the instability and dynamism of natural objects – animate and inanimate (Davis 2010, 564). His aesthetics of natural beauty do not adhere to a humanist paradigm of an artist shaping the natural world in his or her image; instead, form is the outcome of inherent temporal forces, or *poiesis* (Davis 2010, 572). Ultimately, Thoreau's environmental ethos led him to reject aesthetic framing in terms that would have been familiar to Gilpin and other painters of the picturesque. A critique of the ornament is evident in *Wild Fruits* when he asks, "what, for instance, are the blue juniper berries in the pasture, considered as mere objects of beauty, to church or state?" (Thoreau 2000, 5). The visual beauty of the berries as ornaments is aligned with the dogma of church and state – those twin foundations of American democracy. In contrast, the sensuous and edible attributes of the berries embody the obverse: wildness. Whereas an ornament is visual rather than functional (excepting, for example, some architectural ornaments), the blue juniper berries are beautiful (visually appealing), sensuous (edible), and serviceable (used for the production of alcohol). This underscores that fact that Thoreau's embodied aesthetics is concerned with wild plants – those that consort with him in loosening the humanist grip on tenets of beauty defined through art and sight. In many instance, he refers to the "wild flavor" of certain fruits (Thoreau 2000, 59), or those like the wild gooseberry that are "rather acid and wild tasted" (62).

FROM KANT TO THOREAU TO SERRES: RECLAIMING THE SENSE OF TASTE

One of the ways in which Thoreau engenders an aesthetics of flora and thereby rejects the ocularcentric Kantian tradition is through the simple yet radical acts of nibbling, tasting, consuming, and processing as food the berries of the Concord area. The sense of taste, however, has a much beleaguered position in the history of Western aesthetics. Aristotle only recognized four senses, correlating them to the four elements: vision with water, sound with air, smell with fire, and touch with earth. He regarded taste as a derivative of touch (Connor 2008, 2). Later philosophers pejoratively considered the olfactory and gustatory to be base or primordial levels of being (Assad 1999, 84). For German philosopher Immanel Kant (1724–1804), interested in establishing a system of aesthetic judgements based on pure beauty, taste is not the sense itself but a metaphor for aesthetic sensibility in general (i.e. as Taste) (Korsmeyer 1999, 54). What results from Kant's metaphysics is a sense hierarchy, segregating judgements based on the distal senses of vision and hearing from those derived from the proximal senses of pleasure (Korsmeyer 1999, 54).

The sense of taste as gustation, for Kant, entails bodily sensation not free from desire (hence not disinterested) and, therefore, fails to lead to the pure aesthetic judgement of beauty. As base, primordial, and carnal drives, hunger and sexual appetite interrupt pure aesthetic contemplation and the formation of judgements that could be considered valid between people and thus universal (Korsmeyer 1999, 56). Kant distinguishes between the "objective" senses of seeing, hearing, and touch in contrast to the "subjective" senses of smelling and tasting: "The subjective senses are senses of enjoyment, the objective senses, on the other hand, are instructive senses" (Kant quoted in Berger 2009, 34). For Kant, whereas the three objective senses principally (and more consistently) convey information about objects, the two subjective senses lead to highly subjective experiences of pleasure or displeasure. In his *Lectures on Metaphysics*, presented between the 1760s and 1790s, and later published, Kant asserts that "if one merely smells or tastes, one can not yet distinguish one thing from another. I cannot know color, shape, etc. [...] We can fall info a swoon from strong odors, and from foul taste nausea can be aroused and thereby set the entire body into convulsions" (Kant 1997, 251). As the hallmarks of visual beauty, color and shape relate to cognitive knowledge. In sharp contrast, smell and taste can result in negative effects on the body that occur regardless of our conscious faculties.

The contemporary French philosopher Michel Serres (b. 1930) counters the Kantian hierarchy of the senses that largely privileges the distal over the proximal senses – vision and hearing over touch, taste, and smell. For Serres, the intermingling of the senses is the mechanism through which the body interacts with the world and transcends the physical and existential boundaries of human subjectivity (Connor 2008, 3). Serres disturbs the Kantian paradigm by stressing the correspondence between the sense of taste, the attainment of knowledge, and the faculty of language. "What we hear, through our tongue, is that there is nothing in sapience that has not first passed through mouth and taste, through sapidity" (Serres 2008, 162). Sapidity (the quality of having flavor) mirrors sapience (the quality of having wisdom and discernment) – the two words sharing an etymology in the Latin *sapere*, meaning both to taste and to be wise. In other words, both the experience of taste and the enunciation of wisdom (in the form of language) pass through the mouth and involve the tongue as the shared organ (Assad 1999, 84). *Homo sapiens*, then, are beings who both taste and know; or know *through* taste. "Wisdom comes after taste, cannot arise without it, but has forgotten it [...] taste institutes sapience" (Serres 2008, 154). The modes of abstraction and analysis associated with sight and hearing "tear the body to pieces," negating taste, smell, and touch (Serres 2008, 26). The antidote is a "return to things themselves" (Serres 2008, 112), a return to the proximal senses, those which put human experience into direct, unmediated contact with the world and the body. On the contrary, logic and grammar (the tenets of language) become "dreary and insane when they deny themselves bodies" (Serres 2008, 235). A language of taste is necessarily situated in the body; the tongue of taste is the tongue of language. However, the experience of taste is never confined to the physiological actions of the tongue in which taste receptors receive sensations from the substances of food and drink. Instead, using the example of wine, Serres constructs taste as the integration of climatic conditions, soil formations, wind patterns, water conditions, sun angles, and cultivation practices (Assad 1999, 83). Put differently, taste is *a priori* an environmental sense that experientially maps onto its ecological provenance. Its boundaries (which separate it from the other senses) dissolve as its effects intermingle with the environment, the body, and sensation itself.

In *Wild Fruits*, Thoreau expresses this latter aspect of Serres' philosophy of taste eloquently. The taste of the wild-crafted berry embodies the taste of the earth, the environment, the seasons, the soil, the elements, the stars, the wetlands. One of the earliest wild fruits of spring, the strawberry makes possible a gustatory experience of the earth specific to this time of year:

> What flavor can be more agreeable to our palates than that of this little fruit, which thus, as it were, exudes from the earth at the very beginning of the summer, without any care of ours? What beautiful and palatable bread! [...] I taste a little strawberry-flavored earth with them. I get enough to redden my fingers and lips at least. (Thoreau 2000, 12)

This passage disrupts an aesthetics of the ornament (of "this little fruit"), focusing instead on the strawberry-infused taste of earth and the tactility of reddening fingers and lips. Thoreau (2000, 12) likens the strawberry to a "concentration and embodiment of that vernal fragrance with which the air has lately teemed." The condensation of spring's fragrance is both in the image and taste of the strawberries. The fruit as a "palatable bread" reflects Thoreau's knowledge of Native American cultures, particularly the reliance of some societies on pemmican, a dense mixture of fat, protein, and, depending on the season or ceremony, fruits. Moreover, the acidic fruits of high blueberry (*Vaccinium corymbosum*) "embody for me the essence and flavor of the swamp" (Thoreau 2000, 31) with their "little blue sacks full of swampy nectar and ambrosia commingled, whose bonds you burst by the pressure of your teeth" (33). We thus find in Thoreau's aesthetics of flora a distinctive ecological aesthetics of taste in which gustatory experiences of fruit are implicated with the environment in which the fruit matures and from which the fruit extracts a particular local flavor. The Serresian mingling of the senses, in this instance, involves synergism between vision ("little blue sacks"), touch ("bonds you burst"), and taste ("swampy nectar"). Other examples are apparent throughout the text. The early low, or dwarf, blueberry (*Vaccinium pennsylvanicum*) bears "a very innocent ambrosial taste, as if made of the ether itself" (Thoreau 2000, 22). The taste of the fruit is the taste of "ether," from the Latin *aethēr* for pure, bright, rarefied air; and invoking the ancient alchemical element – the fifth, after air, earth, fire, and water – thought to be ubiquitous in the heavens but out of the reach of human perception. In contrast, the fruit of late low blueberry (*Vaccinium vacillans*) is "more like solid food, hard and bread-like, though at the same time more earthy" (36), further revealing Thoreau's elemental ideas concerning plants and their fruits. Finally, the pores of a pear "whisper of the happy stars under whose influence they have grown" (128).

THOREAU'S ECOLOGICAL SENSE OF TASTE: THEMES IN *WILD FRUITS*

Turning from the conceptualization of taste (with a lower case "t") in Kantian and post-Kantian philosophies, this section will analyze the dominant themes in *Wild Fruits* that coalesce Thoreau's ecological gustation. The gustatory philosophy presented in *Wild Fruits* follows Serres' assertion that, rather than an undeveloped, isolated, and merely carnal sense, taste imparts complex knowledge and wisdom; and that taste intermingles with our other senses in our experiences of it and the natural world. To taste is also to smell, touch, hear, see, think, dream, and imagine. For Thoreau, the practice of tasting (or, often, nibbling) fruits is continuously informed by Native American and Anglo-European botanical traditions, both of which are contingent upon largely proximal – rather than entirely distal – interactions with plants satisfying the (unmistakably "interested") carnal drive to consume foods and medicines, to attain nourishing substances, to find relief from disease, and, eventually, to survive and even flourish in one's environment. Conversant with these traditions, Thoreau references key studies along with his personal observations of Native American and Anglo-European ethnobotanies. Through these means, he develops a sophisticated empiricism of taste, which cultivates, rather than mutes, the discriminatory and knowledge-making capacities of this most "subjective" and primal sense. Although Thoreau (quoted in Friesen 1984, 15) at one point characterizes the sense of taste as "commonly gross," he suggests that regular practices of gustation assist in developing human acuteness of perception.

In *Wild Fruits*, taste is not isolated from its manifold sensory, environmental, and cultural contexts. The sensuous aesthetics of the text is ostensibly informed by Native American traditions of harvesting wild foods and, in particular, consuming berries. Lawrence Willson (1959) and, more recently, Timothy Troy (1990) have noted Thoreau's intensive interest in the cultural traditions of Native Americans. Thoreau also made use of what we would today call ethnographic approaches, particularly one-on-one field interviews and "mobile ethnographies" (Evans and Jones 2011) involving walking and other forms of movement, to access environmental knowledge and understand the natural history of the Concord area. In his extended rumination on the black huckleberry, Thoreau observes plainly that "the berries *which I celebrate* [and which most other Anglo-Europeans do not] appear to have a range, most of them, very nearly coterminous with what has been called the Algonquin Family of Indians [...] these were the small fruits of the Algonquin and Iroquois Families [emphasis in original]" (Thoreau 2000, 46). In fact, he derived some of his knowledge of edibles from "walking behind an Indian in Maine and observing that he ate some [berries] which I never thought of tasting before" (46). Thoreau was also an advocate for the use of the Native American names for plants, in lieu of "the very inadequate Greek and Latin or English ones at present used" (50); alongside Latin designations, he presents the common, folk, local, indigenous, and historical names of flora.

However, other aspects of Thoreau's ethnobotanical knowledge were second-hand, as he references, for instance, French explorer Jacques Cartier's observation of indigenous Canadians drying plums for the winter, just as the French did (112). As well as Native American sources, Thoreau draws from Gerard's *The Herball or Generall Historie of Plantes* (1597). Like ethnobotanical traditions, herbal knowledge is based upon proximal interactions of tasting, smelling, and touching plants as medicines. Thoreau commends Gerard's careful, embodied reporting of sensations produced by plants and seems to prefer his accounts of English flora to those of other nineteenth-century botanists and naturalists (Friesen 1984, 70–71). The example of the sweet flag is indicative of the extent of Thoreau's reading, encompassing indigenous, ancient Greek and Roman, and contemporaneous sources. Thoreau quotes Gerard, who explains the esteem that Tartars held for the root: "they will not drink water (which is their usual drink) unless they have just steeped some of this root therein" (8). In the same passage, Thoreau subsequently refers to the nineteenth-century Scottish naturalist and explorer Sir John Richardson's documentation of the Cree name *watchuske-mitsu-in* for sweet flag and its use by Native Americans as a treatment for colic. However, knowledge of the palatability of the "inmost tender leaf" was, at least by the mid-nineteenth-century, preserved among Concord children as the folk knowledge of those who went "a-flagging" (sweet flag harvesting) in the spring (7). Again, in his passage on wild strawberries, Thoreau quotes Gerard, who depicts their taste as "little, thin and waterish, and if they happen to putrify in the stomach, their nourishment is naught" (11).

Subtleties of language and expression reveal Thoreau's discerning between pleasurable, neutral, and repellent tastes. Gustatory variations between the opposite poles of agreeable and disagreeable are expressed in his work. Thoreau's acts of nibbling local plants and forming opinions about their qualities underlie an empiricism of taste, in which, contrary to Kant but affirmative of Serres, information is derived through gustation and knowledge is gained. Indeed, his occasional walking companion Ellery Channing discussed Thoreau's "edible religion" involving sustained devotion to sampling, through taste, nearly every wild plant that he could access (Friesen 1984, 27). The red low blackberry has a "lively acid but pleasant taste, with somewhat of the raspberry's spirit. They both taste and look like a cross between a raspberry and a blackberry" (26). Here, Thoreau contemplates the natural hybridization of the raspberry and blackberry that has resulted in a berry with a "raspberry's spirit" – one in which its taste is tantamount to its visual appearance. This practice of empirical deduction constructs the sense of taste not only in terms of generalized appreciation of nature but for its capacity to inform aesthetic judgements, underpin ecological knowledge, and prompt the differentiation between species according to their gustatory qualities (rather than their visual attributes in the Linnaean genus-species taxonomic model, which largely ignores taste). Regarding the smooth sumac, he notes "that sour-tasting white and creamy incrustation [*sic*] between and on the berries of the smooth sumac, like frostwork. Is it not an exudation? Or is it produced by the bite of an insect?" (125). Taste (sour and creamy) precedes sight (frostwork) and initiates deductive questioning regarding the ecological purpose of the unusual encrustation.

The dynamics between taste, smell, vision, and sensuality more broadly constitute a salient theme in Thoreau's ecological gustation. As such, *Wild Fruits* compellingly illustrates Serres' notion of the mingling of the senses. On the late low blueberry, Thoreau observes that "these almost spicy, lingering clusters of blueberries contrast strangely with the bright leaves" (37). This statement is a surprising instance of synaesthesia in which the sapidity of the berries (their piquancy) is pitted against the visual characteristics of the leaves (their intensity). Usually, tastes are compared to other tastes; sights to other sights; but Thoreau disrupts this kind of experiential correspondence and expectation. The dynamics between the senses sometimes result in an opposition, rather than a contrast or complementarity, as the flavor of the blueberries "prevents our observing their beauty" (33). We find an aesthetics that counters the idea of nature as an ornament or decorative object. Thoreau's immersive sensuality – one can imagine his whole face plunged into the bush, mouth ready, and lips taut to pluck the berries – diverges sharply from the disinterested contemplation of beauty inherent to Kant's aesthetic philosophy. The volatile chemicals of fruits are often smelled before they are tasted or seen in an uncanny inversion of visual order and an interpenetration of the senses. The fruit of a particular wild apple tree has a "peculiarly pleasant bitter tang, not perceived till it is three-quarters tasted. It remains on the tongue. As you eat it, it smells exactly like a squash-bug" (86). The mingling of taste (the apple's "bitter tang") and smell (pungent "like a squash-bug") has much to do with the physiology of smell. Olfaction occurs *orthonasally* (through the nostrils into the nasal cavity itself) and *retronasally* (via the palate within our mouths), the former also being the pathway of taste (Todd 2010, 26).

WILD TASTES AND LOCAL FOODS: THOREAU'S AESTHETIC LESSONS

Much of *Wild Fruits* concerns finding the appropriate and most evocative language to capture, convey, and give voice to experiences of tasting fruits (and a few roots) in prose. Ultimately, Thoreau confers a discourse to taste that involves the human *sensorium* – the sum of a being's perception linking sense experiences together as bodily sensation in a place or bioregion. Thoreau's botanical aesthetics give discourse to the wild tastes of local berries. Their abandonment takes place as Anglo-European palates become increasingly accustomed to cultivated fruits. For instance, some wild fruits are highly astringent and considered unpalatable in large quantities. Referring to chokeberry (*Pyrus arbutifolia*), "I *eat* the high blueberry, but I am also interested in the rich-looking, glossy-black chokeberries, which nobody eats and which bend down the bushes on every side – sweetish berries, with a dry and so choking taste [italics in original]" (66). By late August, the chokeberries have "a sweet and pleasant taste enough at first, but leave a mass of dry pulp in the mouth" (66). These are uncelebrated fruits, their profusion a result of their disregard – their disregard a reflection of their caustic flavor and its unsettling physical sensations. This effect is also evident with the choke cherry (*Cerasus virginiana*), which "[so furs] the mouth that the tongue will cleave to the roof, and the throat wax hoarse with swallowing those red bullies" (72). Using an apt domestic metaphor for their astringency, Thoreau observes that "the juice of those taken into the mouth mixed with the saliva is feathered, like tea into which sour milk has been poured" (72). However, this "natural raciness" could have less to do with the inherent qualities of the fruits themselves and more to do with their reception, as "the sours and bitters which the *diseased* palate refuses, are the true condiments [italics added]" (87).

Other variations of wild taste are more pleasurable and desirable than their cultivated counterparts, at least to Thoreau's palate. With an air of regional and national solidarity, Thoreau emphasizes that these flavors distinguish the Concord (and, more broadly, the American) landscape. The taste of wild apples is "more memorable [...] than the grafted kinds; more racy and wild American flavors do they possess [...] an old farmer in my neighborhood, who always selects the right word, says that 'they have a kind of bow-arrow tang'" (2000, 84). Moreover, the apple's flavor is contingent on its environment and dramatically transforms for the worse when brought indoors, that is, as it becomes domesticated. The fruit, "so spirited and racy when eaten in the fields or woods, being brought into the house, has frequently a harsh and crabbed taste" (85). Here, we find a friction between the tastes of domestic (or cultivated) and wild (or uncultivated) fruits, the latter needing to be consumed in the environment in which it matured in order to be fully appreciated. Thoreau explains that wild flavors are designed to be savored in their natural settings and with same freedom of spirit exerted during their collection: "the Saunterer's Apple not even the saunterer can eat in the house. The palate rejects it there, as it does haws and acorns, and demands a tamed one" (85).

Thoreau's botanical aesthetics, as enunciated in *Wild Fruits*, constitute a proto-bioregional position (bioregionalism being inaugurated as an environmental movement in the Western United States in the 1970s), in which the seasonal foods local to one's region are valued over those imported from elsewhere(for example, see Berg 2014). Indeed, food is a crucial aspect of environmental sustainability (Oosterveer and Sonnenfeld 2012). Thoreau entreats us to consider the merits and pleasures of local consumption. Coming to know the wild fruits of one's area also involves becoming physically immersed and sensuously interconnected. He explains that "our diet, like that of the birds, must answer to the season" (107). Regardless of the effects of modernization and industrialization on food production and consumption, "it is surely better to take thus what Nature offers in her season, like a robin, than to buy an extra dinner" (142). While some wild flavors, such as chokeberries and choke cherries, require the re-education of the human senses to appreciate, others are immediately pleasing and without parallel: "No tarts that I ever tasted at any table possessed such a refreshing, cheering, encouraging acid that literally put the heart in you and set you on edge for this world's experiences, bracing the spirit, as the cranberries I have plucked in the meadows in the spring" (106). As Thoreau dismantles the distance between himself (as subject) and plants (as objects), he at the same time reveals the complexity of the taste faculty and the fruits it promises for a more sustainable, sensual, and posthumanist future.

Nature Writing for the Love of Plants and Place: A Posthuman Practice

Bearing in mind Thoreau's work (chapter 12), as well as other environmental writings included in *Posthuman Plants*, the Epilogue contemplates nature writing as the expression the permeability between people and nonhumans, other-than-humans, and more-than-humans (pick your term): wildflowers, whales, peregrines, and all shapes and sizes of animate and inanimate things between. Such writing can show what nature is to us: not merely a backdrop for human affairs but rather a place of dreams and drama, inspiration and excitation. Nature has a voice to share (as chapter 10 argues). If we listen closely, we can hear its susurrating wisdom.

****

In 1967, J.A. Baker published an evocative account of his observations of peregrine falcons over the course of an English winter. With voyeuristic obsession, he recorded the ritualistic behaviours of peregrines. His fascination is apparent in the halting passages of the book: "Wherever he goes, this winter, I will follow him [...] till my predatory human shape no longer darkens in terror the shaken kaleidoscope of colour that stains the deep fovea of his brilliant eye. My pagan head shall sink into the winter land, and there be purified" (Baker 2011). Baker's *The Peregrine* is meaningful to us still because of its devotion to the more-than-human world; its diligence in exploring landscape cycles; its empathy for the great efforts of wild organisms to survive; and its abiding faith in personal transformation through nature.

Indeed, Baker had to become a devotee of the peregrine. His work suggests that writing can affect our relationships to the living earth through the power of attentiveness. The term *nature writing* originated in the West with the works of Thoreau whose renowned *Walden*, originally published in 1854, narrates his experiment in living closely with Walden Pond (Thoreau 2004). Like Baker, Thoreau attended to the world around him, its pungent tastes and subtle sounds, despite modernity's encroaching distractions and disruptions. Nature writing expresses these sorts of multifaceted relationships between people and more-than-humans. Including novels, poetry, essays, plays, blogs, and emerging multi-media forms, nature writing is writing that addresses the natural world seriously and recognizes nature's capacity for address (see chapters 10 and 11). John Murray (1995, x) in his *Nature Writing Handbook* encapsulates nature writing simply as "literary works that take nature as a theme." Thus, writing is central to an understanding of plants in a posthumanist framework.

For Baker and Thoreau, attentiveness is central – but it is sensory, embodied, visceral enmeshing between humans and more-than-humans that distinguishes nature writing from fact-driven accounts of the environment. As we feel, see, taste, smell, and listen to nature, we become part of what we usually see as a distant scene. Minds quieten and internal analytical chatter softens. The process of writing can be demanding, energetic, and enveloping – as Baker's fidelity attests. In extraordinary instances, nature writing can become spiritual entrainment to the breathing world surrounding us. But the writer's hard work is rewarded by moments of exhilarating insight into the mysteries of birds, plants, and water.

I emphasize here that nature writing operates in a liminal space between culture and nature, between the world and the self. Nature writers are in an uncommon position; they can amble freely in the ecotones between science and poetry, intuition and reason, joy and despondency, metaphors and facts, society and silence. They tend to be generalists with a love of all forms of knowledge. Their writings hybridize – naturally of course – with the world in which they participate. For David Abram, author of *The Spell of the Sensuous*, nature writing entails dialogic exchange between people and the world: "Writing, like human language, is engendered not only within the human community but between the human community and the animate landscape, born of the interplay and contact between the human and the more-than-human world" (Abram 1996, 95).

Hence, writing is also an important medium for *restoring* sensuousness to our relationships with the natural world. Cultural theorist Rod Giblett understands nature writing as "the creative, written tracing of the bodily and sensory enjoyment of both the processes and places of nature" (Giblett 2011, 26). Indeed, seminal nature writing involves a dialogue with the many senses. Expressed as a polyphony, the voice is that of nature and people participating in the land's rhythms. In 1862, Thoreau asked famously, "where is the literature which gives expression to Nature?" For him, the nature writer's work is the practical business of nailing "words to their primitive senses, as farmers drive down stakes in the spring, which the frost has heaved" (Thoreau 1862/2007, 29).

An important theme of nature writing is seasonality. The ecocritic Lawrence Buell (1995, 220) observes that "in environmental prose, the seasons have been a particularly favored organizing principle." Many Western works of nature writing, including Thoreau's *Walden*, Aldo Leopold's *A Sand County Almanac*, and Annie Dillard's *Pilgrim at Tinker Creek*, are seasonal in character and design. Nature writing of the northern hemisphere attends to the changes marking the seasons: the onset of snow, the falling of leaves, and the melting of the ground. In traditional Japanese poetry, a *saijiki* is a dictionary of seasonal words used by poets of traditional forms, like haiku (Ryan 2013, chapter 1).

Today's nature writers should attend to the seasons – and more importantly Indigenous seasonality – in telling a story of the land. The six-season calendar of the Nyoongar, the indigenous people of the South-West corner of Western Australia, is an illuminating example. But each region of Australia has an ecological calendar, reflecting seasonal nuances. The Nyoongar seasons are *djeran, makuru, djilba, kambarang, birok*, and *bunuru* (Ryan 2014, 11–15). As the season of birth, kambarang is marked by a profusion of wildflowers and a decrease in rain. The *wonil* or Peppermint tree (*Agonis flexuosa*) effuses aromatic oils in the hot air during this season. Birok is the hot and dry season of the young, or the "first summer." The mudja, or Christmas tree (*Nuytsia floribunda*), flowers during birok (see chapter 7). The traditional Australian seasons express the voice of the land that can figure into writings.

In the South-West corner of WA, environmental prose and poetry, in printed and digital forms, are poised to grow exponentially as more is known about the ecology of the landscape (see chapters 3 and 4). One notable precedent is *Between Wodjil and Tor*, a close study of a remnant parcel of Wheatbelt flora (Main 1967). So, in addition to the senses and the seasons, another "instrument" available to an emerging or seasoned nature writer is the field journal, that peculiar melting pot of science, intuition, stories, and sense impressions. Journal writings describe "long-term processes of nature, such as seasonal or environmental changes, in great detail," as Murray (1995, 2) says. Writing about place through careful observation and devoted interaction over time has been exemplified more recently in the South-West by *The Lake's Apprentice* (Weldon 2014), the subject of which is Lake Clifton, south of Perth.

The particular vigor of nature writing is its capacity to articulate our relationships to the earth in an era of great pressures on wild places. Walden Pond, near Boston, is now a symbol of the resurrection of nature in American literature. Even after Rachel Carson's warnings in *Silent Spring* in the 1960s and the haunting realities of climate change in the past decades, Thoreau's prose resonates. Enduring nature writing represents landscapes not merely as backdrops for the activities of people but rather as sites of ecological dramas, of interest for their own sake and with a voice to share.

****

A frenzy of lightning streaks across my drowsy mind. What is real mixes with what is imaginary in a hazy zone between midnight and morning. All the cells of my body hum as I lie cocooned in a down-feather sleeping bag. Unsteady rain patters against the taut nylon of my shelter. Half conscious, I hear muffled scratching beside me. I open the noisy zipper of my portable home and peer outside, aiming a small torch at a trembling shape.

With a long muzzle, an echidna rummages through the gnarled and exposed roots of gum trees. Her quills are drawn like needles into the air as she scuffs around for an appetizer to satisfy the sting of hunger. Lapsing into dreaminess again, I imagine the spiny creature burrowing into my belly, crawling into my body, tunnelling around, seeing what could be found *inside me*.

My eyes open. I wait for myself to come back from the nocturnal encounter. The spring sun glows through the tent's translucent walls. I unfurl myself as a morning flower does, like a colourful blossom in my outdoor gear. The dusky clouds clear. Crystalline blueness returns to the sky. Bursts of chilly air carry sharp, fragrant eucalypt scents to my cheeks and nose. I am awakened. Morning smooths out a stormy night's tousling.

I am sleeping in the *kwongan*. Located roughly between Geraldton and Perth in WA, this landscape is a creative miracle. More kinds of plants exist here than in tropical jungles. But one could not estimate the botanical sumptuousness of the *kwongan* from a distance. The low-growing bush survives harsh heat and intense wind because of its small flowers and spiky leaves. I attend sensuously, crawl on my hands and knees, lap drops of wildflower nectar, put my nose into secret nooks and crannies. I adapt too.

On my feet, I search for nerved hakea. It flowers in soft pink tassels during a brief window in late August. I spot a group growing at the edge of paddocks. Lying on the ground again, I look upward through the leaves of the shrub. The nerve-like branching of its veins traces a pattern of lightning across the sky. Was this the electricity I dreamt of last night? How can a plant communicate through sensations in my dream world? I jot down in my field journal a string of fast-flying phrases. My notebook becomes a unique melting pot of science, intuition, stories, and sense impressions that later become poems.   I must admit that I am a botanophilist: a lover of plants. I write poems about and contemplate the existence of wildflowers. To me, trees, shrubs, orchids, and other members of the green world are sacred and intelligent (see chapter 11). In this part of Australia, they also vibrate to very old rhythms. But when I first came to this special part of the world, I felt bewildered. The plant life of Western Australia is remarkably different to the eastern United States with which I was intimately familiar. I grew up in coastal New Jersey, then moved to the lush rolling hills of Massachusetts. For many years, I took walking pilgrimages through the wildernesses of California and Canada. The comfortable sights of oaks and pines have been supplanted with the images and aromas of banksias and tingle trees.

Three weeks on, it is in the middle of September. Spring time. The Southern Ocean foams and pounds furiously against the dramatically dropping coastline. I am on a walk with the Friends of the Fitzgerald River National Park. We ascend sandy undulations, looking behind us from time to time for a glimpse of the sea. Orchid flowers emerge from cracks in the clean grey granite. Their orange and red hues pulsate like panting tongues in the wind. At the top of a plateau, we find another unusual plant: the cousin of the nerved hakea of my electrified dream. Royal hakea was named by early botanists for the glowing orange splotches adorning the prickly crowns of its leaves. Like a tower of stacked cabbage heads, it veers in the breeze, almost topples but springs back to a standing position.

Tentative and shy, I am not sure how to greet this plant. What is the etiquette in the bush? Shaking hands won't do. Looking around furtively for fellow bush walkers, I drop to the ground and crawl. On my back again, I gaze up to the clear Australian heavens. The patterns of veins in each ornate leaf seem to fracture the sky. I have visions of sharks in the sea bearing their teeth. I think of the sharp claws of a primordial animal deep below the ocean's surface. Phrases come naturally and easily. I let them. I write these in my field notebook. The spontaneous flow of writing will eventually become part of a poem about this ragtag colony of plants.

Here is the theme I wish to close *Posthuman Plants* on. Writing is connected to sustainability. While they are stunning and inspiring, these magnificent life forms are also imperilled. As prominent biologists report, we are living in the "sixth great extinction," an era of crushing biodiversity loss (see chapter 1). Commentators estimate that a high percentage of the planet's species might disappear by the end of the twenty-first century. We're realizing, hopefully not too late, that the sustainability of our increasingly global culture is shared with birds, rocks, and flowers. I can only drop to the ground again when I think about the facts. A sobering thought is that only three percent of the original *kwongan* ecosystem remains.

In times of grief and disconnection, nature writing reminds us of our intrinsic connectedness. Our viability as a culture, increasingly globalized, is shared with all other beings. The fate of hakeas is also ours. The right words, those from the heart, can create intimacy and restore sensuousness to our relationships with the world. *Posthuman Plants* has been drawn from a botanophilist's love of plants, my trust in the vegetal and its liberation from the margins and from anthropocentric presumptions. Instead, you could take a concerted interest in animals, rocks, seaweed, the sky, the stars, or the natural world as a breathing whole. Attentiveness can be powerful and comes in many shapes and sizes.

Remarkable places with astonishing plants are in need of recognition, celebration, and protection. Writing and the arts more generally are integral to the appreciation of our intersections. Words, images, and sensations can inspire others. Such expressions can fulfill a private need for connection or mourning. Or they can record experiences of the natural world for our grandchildren and our communities.

References

Abram, David. 1996. *The Spell of the Sensuous: Perception and Language in a More-Than-Human World*. New York: Pantheon Books.

Albrecht, Glenn. 2005. "'Solastalgia': A New Concept in Health and Identity." *PAN: Philosophy, Activism, Nature* 3: 44–59.

Albrecht, Glenn. 2010. "Solastalgia and the Creation of New Ways of Living." In *Nature and Culture: Rebuilding Lost Connections*, edited by Sarah Pilgrim and Jules Pretty, 217–234. Hoboken: Taylor and Francis.

Albrecht, Glenn. 2016. "Solastalgia." In *Mourning Nature: Hope at the Heart of Ecological Loss and Grief*, edited by Ashlee Cunsolo Willox and Karen Landman. Montreal: McGill-Queen's University Press.

Albrecht, Glenn, Gina-Maree Sartore, Linda Connor, Nick Higginbotham, Sonia Freeman, Brian Kelly, Helen Stain, Anne Tonna, and Georgia Pollard. 2007. "Solastalgia: The Distress Caused by Environmental Change." *Australasian Psychiatry* 15 (Supplement): S95–S98. doi: 10.1080/10398560701701288.

Alcampo, Jo SiMalaya. 2010. *InterAccess* Retrieved June 5, 2014, from www.singingplants.org/images.html.

Alexander, Alan. 1979. "Nuytsia Floribunda." In *Wide Domain: Western Australian Themes and Images*, edited by Bruce Bennett and William Grono, 53–54. London: Angus & Robertson.

Alexander, Bryan. 2011. *The New Digital Storytelling: Creating Narratives with New Media*. Santa Barbara: Praeger.

American Cancer Society. 2008. *Pokeweed* Retrieved September 5, 2014, from http://www.cancer.org/treatment/treatmentsandsideeffects/complementar yandalternativemedicine/herbsvitaminsandminerals/pokeweed.

Appelbaum, David. 1990. *Voice*. Albany, NY: State University of New York.

Ascott, Roy. 2000. "Edge-Life: Technoetic Structures and Moist Media." In *Art, Technology, Consciousness: mind@large*, edited by Roy Ascott, 2–6. Bristol: Intellect.

Ashbolt, Paul, Benjamin Quaife, and Sarah Ryan-Charles. 2012. "Phenological Aspects of *Nuytsia floribunda*." *Cygnus* 1: 207– 217.

Assad, Maria L. 1999. *Reading with Michel Serres: An Encounter with Time*. Albany, NY: State University of New York.

Ayris, Cyril. 2001. *Araluen: Valley of Dreams*. West Perth: Cyril Ayris Freelance.

Baker, J.A. 2011. *The Peregrine: The Hill of Summer & Diaries: The Complete Works of J.A. Baker*. London: HarperCollins.

Balch, Phyllis. 2002. *Prescription for Herbal Healing*. New York: Avery.

Balick, Michael, and Paul Cox. 1996. *Plants, People, and Culture: The Science of Ethnobotany*. New York: Scientific American Library.

Bannister, Thomas. 1833. "A Report of Captain Bannister's Journey to King George's Sound, over Land." In *Journals of Several Expeditions Made in Western Australia During the Years 1829, 1830, 1831, and 1832*, edited by J. Cross, 98–109. London: J. Cross.

Barad, Karen. 2010. "Quantum Entanglements and Hauntological Relations of Inheritance: Dis/continuities, SpaceTime Enfoldings, and Justice-to-Come." *Derrida Today* 3 (2): 240– 268.

Barker, R.M. 1996. "James Drummond's Newspaper Accounts of his Collecting Activities, in Particular his 4th Collection and *Hakea victoria* (Proteaceae)." *Nuytsia* 11 (1): 1–9.

Barlow, Peter. 2006. "Charles Darwin and the Plant Root Apex: Closing a Gap in Living Systems Theory as Applied to Plants." In *Communication in Plants: Neuronal Aspects of Plant Life*, edited by František Baluška, Stefano Mancuso and Dieter Volkmann, 37–51. Berlin, Germany: Springer-Verlag.

Barras, Colin. 2008. *The Green Revolution: Plants Move Online* Retrieved March 11, 2015, from, http://www.newscientist.com/blogs/shortsharpscience/2008/10/the-japanese-have-opened-up.html.

Barthes, Roland. 1977. *Image, Music, Text*. Translated by Stephen Heath. London: Fontana Press.

Bate, Jonathan. 2000. *The Song of the Earth*. Cambridge, MA: Harvard University Press.

Bates, Daisy. 1992. *Aboriginal Perth and Bibbulmun Biographies and Legends*. Edited by P.J. Bridge. Victoria Park, WA: Hesperian Press.

Beard, J. 1979. "Phytogeographic Regions." In *Western Landscapes*, edited by Joseph Gentilli, 107–121. Nedlands: University of Western Australia Press.

Becker, Lawrence. 2014. *Reciprocity*. Oxon: Routledge and Kegan Paul Ltd.

Benjamin, Walter. 1996. "On Language as Such and on the Language of Man." In
        Selected Writings: 1913–1926, Volume 1,edited by M. Bullock and
        M.W. Jennings, 62–74. Cambridge, MA: Belknap Press.

Beresford, Quentin, Hugo Bekle, Harry Phillips, and Jane Mulcock. 2001. *The
        Salinity Crisis: Landscape, Communities and Politics* Crawley:
        University of Western Australia Press.

Berg, Peter. 2014. *The Biosphere and the Bioregion: Essential Writings of Peter
        Berg*. Edited by Cheryll Glotfelty and Eve Quesnel. Milton Park:
        Routledge.

Berger, David. 2009. *Kant's Aesthetic Theory: The Beautiful and Agreeable*.
        London: Continuum.

Berleant, Arnold. 1991. *Art and Engagement*. Philadelphia: Temple University
        Press.

Berleant, Arnold. n.d. *Thoreau's Aesthetics of Nature* Retrieved April 30, 2015,
        fromhttp://www.autograff.com/berleant/pages/Thoreau%27s%20Aesthet
        ics%20of%20Nature%20.6.htm.

Blau, Tatjana, and Mirabai Blau. 2003. *Buddhist Symbols*. New York: Sterling
        Publishing.

Bletsoe, Elisabeth. 2010. *Pharmacopoeia and Early Selected Works*. Exeter:
        Shearsman Books.

Blunt, Wilfrid, and William Stearn. 1950. *The Art of Botanical Illustration*. Kew,
        England: Antique Collectors' Club in association with the Royal Botanic
        Gardens.

Bock, Jannika. 2008. *Concord in Massachusetts, Discord in the World: The
        Writings of Henry Thoreau and John Cage*. Frankfurt: Peter Lang.

Bondi, Liz, Joyce Davidson, and Mick Smith. 2005. "Introduction: Geography's
        'Emotional Turn'." In *Emotional Geographies*, edited by Joyce
        Davidson, Liz Bondi and Mick Smith, 1–16. Aldershot, England:
        Ashgate.

Brabham, D. 2013. "Crowdsourcing: A Model for Leveraging Online
        Communities." In *The Participatory Cultures Handbook*, edited by A
        Delwiche and J Henderson, 120–129. Abingdon: Taylor and Francis.

Breeden, Stanley, and Kaisa Breeden. 2010. *Wildflower Country: Discovering
        Biodiversity in Australia's Southwest*. Fremantle, Australia: Fremantle
        Press.

Brisbois, Michael. 2005. "Tolkien's Imaginary Nature: An Analysis of the Structure of Middle-earth." *Tolkien Studies* 2: 197–216.

Bristow, Tom. 2008. "Ecopoetics." In *The Facts on File Companion to World Poetry: 1900 to the Present*, edited by R.V. Arana, 156–159. New York: Infobase Publishing.

Brown, Deidre. 2007. "*Te Ahu Hiko*: Digital Cultural Heritage and Indigenous Objects, People, and Environments." In *Theorizing Digital Cultural Heritage: A Critical Discourse*, edited by Fiona Cameron and Sarah Kenderdine, 77–92. Cambridge, MA: MIT Press.

Bryson, J. Scott. 2002. "Introduction." In *Ecopoetry: A Critical Introduction*, edited by J. Scott Bryson, 1–13. Salt Lake City: The University of Utah Press.

Buddhadasa, Bhikkhu. 1989. *Me and Mine: Selected Essays of Bhikkhu Buddhadasa*. Albany, NY: SUNY Press.

Buell, Lawrence. 1995. *The Environmental Imagination: Thoreau, Nature Writing, and the Formation of American Culture*. Cambridge, MA: Harvard University Press.

Butler, Judith. 2004. *Precarious Life: The Powers of Mourning and Violence*. London: Verso.

Callicott, J. Baird. 1993. "The Conceptual Foundations of the Land Ethic." In *Environmental Philosophy: From Animal Rights to Radical Ecology*, edited by Michael Zimmerman and J. Baird Callicott, 110–134. Englewood Cliffs: Prentice-Hall.

Cameron, Fiona, and Sarah Kenderdine. 2007. "Introduction." In *Theorizing Digital Cultural Heritage: A Critical Discourse*, edited by Fiona Cameron and Sarah Kenderdine, 1–15. Cambridge, MA: MIT Press.

Cancer Research UK. 2013. *DSA from Jimson Weed for Brain Tumours* Retrieved September 5, 2014, from, http://www.cancerresearchuk.org/cancer-help/about-cancer/cancer-questions/dsa-from-jimson-weed-for-brain-tumours.

Carbaugh, Donal. 2014. "Response Essay: Environmental Voices Including Dialogue with Nature, Within and Beyond Language." In *Voice and Environmental Communication*, edited by Jennifer Peeples and Stephen Depoe, 241–256. Palgrave Macmillan.

Carbon Arts. n.d. *Melbourne Mussel Choir* Retrieved October 18, 2013, from, http://www.carbonarts.org/projects/melbourne-mussel-choir/.

Carpenter, Humphrey. 1977. *J. R. R. Tolkien: A Biography*. Boston: George Allen
    and Unwin.
Chamovitz, Daniel. 2012. *What a Plant Knows: A Field Guide to the Senses*. New
    York: Scientific American/ Farrar, Straus and Giroux.
Christoff, Peter, and Robyn Eckersley. 2013. *Globalization and the Environment*.
    Plymouth: Rowman & Littlefield Publishers.
Clark, Timothy. 2011. *The Cambridge Introduction to Literature and the
    Environment*. Cambridge, UK: Cambridge University Press.
Clarke, Marcus. 1993. "Preface to Adam Lindsey Gordon's *Sea Spray and Smoke
    Drift*." In *The Penguin Book of 19th Century Australian Literature*,
    edited by Michael Ackland, 43–46. Ringwood: Penguin Books.
Clarke, Philip. 2007. *Aboriginal People and their Plants*. Dural Delivery Centre:
    Rosenberg.
Clewell, Tammy. 2002. Mourning Beyond Melancholia: Freud's Psychoanalysis
    of Loss. *Journal of the American Psychoanalytic Association* 52 (1): 43–
    67, http://www.apsa.org/portals/1/docs/japa/521/clewell.pdf.
Connor, Steven. 2008. "Introduction." In *The Five Senses: A Philosophy of
    Mingled Bodies*, 1–16. London: Continuum.
Conron, John. 2000. *American Picturesque*. University Park, PA: The
    Pennsylvania State University.
Conservation International. 2008. "Biological diversity in Southwest Australia."
    In *Encyclopedia of Earth*, ed. Kevin Caley. Washington, DC:
    Environmental Information Coalition, National Council for Science and
    the Environment. Retrieved April 5, 2011, from,
    http://www.eoearth.org/article/Biological_diversity_in_Southwest_Austr
    alia
Conservation International. 2013. *Southwest Australia* Retrieved April 30, 2013,
    from, www.conservation.org/where/priority_areas/hotspots/asia-
    pacific/Southwest-Australia/Pages/default.aspx.
Convery, Ian, and Peter Davis, eds. forthcoming. *Shifting Interpretations of
    Natural Heritage*. Woodbridge, UK: Boydell Press.
Cooper, Laurence. 1999. *Rousseau, Nature, and the Problem of the Good Life*.
    University Park: Pennsylvania State University Press.
Cordis. 2008. *Midori-san or the Slightly Potty Blog* Retrieved March 11, 2015,
    from,http://cordis.europa.eu/express/20081024/finally_en.html.

Crabbe, Katharyn. 2000. "*The Quest as Legend*: The Lord of the Rings." In *Modern Critical Interpretations: J.R.R. Tolkien's The Lord of the Rings*, edited by Harold Bloom, 141–170. Philadelphia: Chelsea House Publishers.

Croom, Edward. 1992. "Herbal Medicine Among the Lumbee Indians." In *Herbal and Magical Medicine*, edited by James Kirkland, Holly Mathews, Charles Sullivan and Karen Baldwin, 137–169. Durham: Duke University Press.

Csordas, Thomas. 1989. "The Sore That Does Not Heal: Cause and Concept in the Navajo Experience of Cancer." *Journal of Anthropological Research* 45 (4): 457–485.

Davis, Theo. 2010. "'Just Apply a Weight': Thoreau and the Aesthetics of Ornament." *ELH* 77 (3): 561-587. doi: 10.1353/elh.2010.009.

Dean, Bradley P. 2000a. "Introduction." In *Wild Fruits: Thoreau's Rediscovered Last Manuscript*, edited by Bradley P. Dean, ix–xvii. New York: W. W. Norton & Company.

Dean, Bradley P. 2000b. "A Thoreau Chronology." In *Wild Fruits: Thoreau's Rediscovered Last Manuscript*, edited by Bradley P. Dean, 273–275. New York: W. W. Norton & Company.

DeLappe, Joseph. 2011. *Iraqimemorial.org: Commemorating Civilian Deaths* Retrieved April 8, 2015, from,http://iraqimemorial.org/.

DeLappe, Joseph. 2013a. *About the Project* Retrieved October 18, 2013, from, http://www.project929.com/.

DeLappe, Joseph. 2013b. *Project 929: Mapping the Solar* Retrieved October 17, 2013, from, http://rhizome.org/profiles/josephdelappe/.

Deleuze, Gilles, and Félix Guattari. 2004. *A Thousand Plateaus*. London: Continuum.

Demos, T.J. 2010. "Globalization and (Contemporary) Art." In *Art and Globalization*, edited by James Elkins, Zhivka Valiavicharska and Alice Kim, 209–213. University Park: The Pennsylvania State University Press.

Derrida, Jacques, and Eric Prenowitz. 1995. "Archive Fever: A Freudian Impression." *Diacritics* 25 (2): 9–63.

Deussen, Oliver, and Bernd Lintermann. 2005. *Digital Design of Nature: Computer Generated Plants and Organics*. Translated by Anna Dowden-Williams. Berlin: Springer.

Dickerson, Matthew, and Jonathan Evans. 2006. *Ents, Elves, and Eriador: The Environmental Vision of J. R. R. Tolkien*. Lexington, KY: University Press of Kentucky.

Diehl, Joanne Feit. 2005. "Introduction." In *On Louise Glück: Change What You See*, edited by Joanne Feit Diehl, 1–22. Ann Arbor, MI: University of Michigan.

Dillard, Daniel. 2013. "'What is Man but a Mass of Thawing Clay?': Thoreau, Embodiment, and the Nineteenth-Century Posthuman." *Journal for the Academic Study of Religion* 26 (3): 254–269.

DiMaggio, Paul, and Michael Useem. 1989. "Cultural Democracy in a Period of Cultural Expansion: The Social Composition of Arts Audiences in the United States." In *Art and Society: Readings in the Sociology of the Arts*, edited by Arnold Foster and Judith Blau, 141–175. Albany, NY: State University of New York Press.

Dimmitt, Mark. 2000. "Flowering Plants of the Sonoran Desert." In *A Natural History of the Sonoran Desert*, edited by Steven Phillips and Patricia Wentworth Comus, 153–264. Tucson, AZ: Arizona-Sonora Desert Museum Press/ University of California Press.

Dioscorides Pedanius (of Anazarbos.). 2005. *De Materia Medica*. Translated by L. Y. Beck. Hildesheim: Olms-Weidmann.

Dodd, Elizabeth. 1992. *The Veiled Mirror and the Woman Poet: H.D., Louise Bogan, Elizabeth Bishop, and Louise Glück*. Columbia: University of Missouri Press.

Dodds, Joseph. 2011. *Psychoanalysis and Ecology at the Edge of Chaos: Complexity Theory, Deleuze/Guattari and Psychoanalysis for a Climate in Crisis*. Hoboken: Taylor and Francis.

Dominion Herbal College. 1969. *Chartered Herbalist Course Book Two*. Burnaby, BC: Dominion Herbal College.

Dorfman, E. 2011. *Intangible Natural Heritage: New Perspectives on Natural Objects*. Hoboken: Taylor & Francis.

Drucker, Johanna. 2009. *SpecLab: Digital Aesthetics and Projects in Speculative Computing*. Chicago: University of Chicago Press.

Eisenberg, David, Roger Davis, Susan Ettner, Scott Appel, Sonja Wilkey, Maria Van Rompay, and Ronald Kessler. 1998. "Trends in Alternative Medicine Use in the United States, 1990–1997: Results of a Follow-Up

National Survey." *The Journal of the American Medical Association* 280:1569–1575.

Esposito, Chiara. 2013. *The Dream of Flying* Retrieved February 15, 2014, from, www.interface.ufg.ac.at/blog/the-dream-of-flying/.

Evans, James, and Phil Jones. 2011. "The Walking Interview: Methodology, Mobility and Place." *Applied Geography* 31: 849–858.

Evelyn, John. 1699. *Kalendarium Hortense or, The Gard'ners Almanac*. London: George Haddleston.

Faludi, Rob. 2013. *New York Times on Botanicalls, Again!* Retrieved March 11, 2015, from, http://www.botanicalls.com/.

Farr, Douglas. 2012. *Sustainable Urbanism: Urban Design with Nature*. Hoboken, NJ: John Wiley & Sons.

Feist, Richard, Chantal Beauvais, and Rajesh Shukla. 2010. "Introduction." In *Technology and the Changing Face of Humanity*, edited by Richard Feist, Chantal Beauvais and Rajesh Shukla, 1–21. Ottawa: University of Ottawa Press.

Fenves, Peter. 1996. "The Genesis of Judgment: Spatiality, Analogy, and Metaphor in Benjamin's "On Language as Such and on Human Language"." In *Walter Benjamin: Theoretical Questions*, edited by David Ferris, 75–93. Stanford, CA: Stanford University Press.

Ferlay, J, I Soerjomataram, M Ervik, R Dikshit, S Eser, C Mathers, M Rebelo, D Parkin, D Forman, and F Bray. 2013. GLOBOCAN 2012 v1.0, Cancer Incidence and Mortality Worldwide. *IARC CancerBase*, http://globocan.iarc.fr/Pages/fact_sheets_population.aspx.

Field, Barron. 1825. *Geographical Memoirs on New South Wales*. London: John Murray.

Fields, Rick. 1992. *How the Swans Came to the Lake: A Narrative History of Buddhism in America*. Boston: Shambhala.

Fletcher, Kim. 2013. Unpublished Interview with Kim Fletcher by John Ryan. Perth, Western Australia, April 2013.

Flieger, Verlyn. 2000. "Taking the Part of Trees: Eco-Conflict in Middle-earth." In *J. R. R. Tolkien and His Literary Resonances: Views of Middle-earth*, edited by George Clark and Daniel Timmons, 147–158. Westport, CT: Greenwood Press.

Fontham, Elizabeth, and Su Joseph. 2005. "Prevention of Cancers of the Esophagus and Stomach." In *Preventative Nutrition: The Comprehensive*

Guide for Health Professionals, edited by Adrianne Bendich and
Richard Deckelbaum, 25–54. Totowa: Human Press.

Fowler, Jeaneane. 1999. *Humanism: Beliefs and Practices*. Brighton: Sussex
Academic Press.

Francis, Leslie. 2009. "Reciprocity and Environmental Obligations." *Hofstra Law
Review* 37: 1007–1014.

Freud, Sigmund. 1968. "Mourning and Melancholia." In *The Standard Edition of
the Complete Psychological Works of Sigmund Freud, Volume XIV
(1914–1916)*, edited by James Strachey, 243–258. London: The Hogarth
Press and the Institute of Psycho-Analysis.

Friedlander, Larry. 2008. "Narrative Strategies in a Digital Age: Authorship and
Authority." In *Digital Storytelling, Mediatized Stories: Self-
Representations in New Media*, edited by Knut Lundby, 177–195. New
York: Peter Lang.

Friesen, Victor Carl. 1984. *The Spirit of the Huckleberry: Sensuousness in Henry
Thoreau*. Edmonton: The University of Alberta Press.

Gach, Gary, ed. 1998. *What Book?!: Buddha Poems from Beat to Hiphop*.
Berkeley, CA: Parallax Press.

Gagliano, Monica. 2012. "Green Symphonies: A Call for Studies on Acoustic
Communication in Plants." *Behavioral Ecology*. doi:
10.1093/beheco/ars206.

Gagliano, Monica. 2013. "The Flowering of Plant Bioacoustics: How and Why?"
Behavioral Ecology 1–2. doi: 10.1093/beheco/art021.

Gagliano, Monica, Stefano Mancuso, and Daniel Robert. 2012. "Towards
Understanding Plant Bioacoustics." *Trends in Plant Science* 1–3.

Gagliano, Monica, Michael Renton, Martial Depczynski, and Stefano Mancuso.
2013. "Experience Teaches Plants to Learn Faster and Forget Slower in
Environments Where It Matters." *Oecologia* 175: 63–72.

Garrard, Greg. 2004. *Ecocriticism*. London: Routledge.

Gerard, John. 1597. *The Herball or General Historie of Plantes*. London: John
Norton.

Giblett, Rod. 1996. *Postmodern Wetlands: Culture, History, Ecology*. Edinburgh,
Scotland: Edinburgh University Press.

Giblett, Rod. 2006. *Forrestdale: People and Place*. Bassendean, Australia:
Access Press.

Giblett, Rod. 2011. *People and Places of Nature and Culture*. Bristol: Intellect Press.

Giblett, Rod, and David James. 2009. "Anstey-Keane Botanical Jewel." *Landscope* 24 (4): 42–44.

Girardet, Herbert. 2008. *Cities, People, Planet: Urban Development and Climate Change*. 2 ed. Chichester, England: John Wiley.

Glück, Louise. 1992. *The Wild Iris*. New York: HarperCollins.

Goddard, Julia Bachope. 1871. *Wonderful Stories from Northern Lands*. London: Spottiswoode and Co.

Goldsmith, Kenneth. 2011. *UbuWeb* Retrieved April 30, 2015, from, http://www.ubu.com/resources/index.html.

Gordon, Maggie. 2002. "A Woman Writing About Nature: Louise Glück and 'the Absence of Intention'." In *Ecopoetry: A Critical Introduction*, edited by J. Scott Bryson, 221–231. Salt Lake City: The University of Utah Press.

Graham, Beryl. 2007. "Redefining Digital Art: Disrupting Borders." In *Theorizing Digital Cultural Heritage: A Critical Discourse*, edited by Fiona Cameron and Sarah Kenderdine, 93–112. Cambridge: The MIT Press.

Graham, Mary. 2008. Some Thoughts about the Philosophical Underpinnings of Aboriginal Worldviews. *Australian Humanities Review* (45), http://www.australianhumanitiesreview.org/archive/Issue-November-2008/graham.html.

Graves, Robert. 1955. *The Greek Myths, Vol 1*. Baltimore: Penguin Books.

Gray, Carole. 1996. *Inquiry through Practice: Developing Appropriate Research Strategies* Retrieved April 9, 2014, from, www.carolegray.net/Papers_PDFs/ngnm.pdf.

Gray, Robert. 1979. *Grass Script*. London: Angus & Robertson Publishers.

Gray, Robert. 1993. *Certain Things*. Melbourne: William Heinemann.

Gray, Robert. 2008. *The Land I Came Through Last*. Artarmon, NSW: Giramondo.

Gray, Robert. 2012. *Cumulus*. St Kilda, Victoria: John Leonard Press.

Greenberg, Jonathan. 2011. *Modernism, Satire and the Novel*. Cambridge, UK: Cambridge University Press.

Gregerson, Linda. 2001. "The Sower Against Gardens." *Kenyon Review* 23 (1): 115–133.

Groys, Boris. 2012. *Under Suspicion: A Phenomenology of Media*. New York: Columbia University Press.

Hall, Matthew. 2011. *Plants as Persons: A Philosophical Botany*. Albany, NY: SUNY Press.

Hall, Stuart. 2001. "Constituting an Archive." *Third Text* 15 (54): 89–92. doi: 10.1080/09528820108576903.

Hamel, Paul, and Mary Chiltoskey. 1975. *Cherokee Plants and Their Uses: A 400 Year History*. Sylva: Herald Publishing Co.

Haraway, Donna. 2008. *When Species Meet*. Minneapolis: University of Minnesota Press.

Haraway, Donna. 2011. SF: Science Fiction, Speculative Fabulation, String Figures, So Far. In *Pilgrim Award Acceptance Speech presented at the Science Fiction Research Association Conference*. Lublin, Poland.

Harmon, David. 2001. "On the Meaning and Moral Imperative of Diversity." In *On Biocultural Diversity: Linking Language, Knowledge, and the Environment*, edited by Luisa Maffi, 53–70. Washington and London: Smithsonian Institution Press.

Harmon, David. 2013. "A Bridge over the Chasm: Finding Ways to Achieve Integrated Natural and Cultural Heritage Conservation." In *Natural Heritage: At the Interface of Nature and Culture*, edited by Peter Howard and Thymio Papayannis, 73–85. Hoboken: Taylor and Francis.

Harper, Douglas. 2014. *Voice* Retrieved April 11, 2015, from, http://www.etymonline.com/index.php?term=voice.

Harpur, Charles. 1973. *Charles Harpur: Book One, Three Colonial Poets*. Edited by Adrian Mitchell. Melbourne: Sun Books.

Hart-Smith, William. 1979. "Kangaroo Paw " *Artlook* 5 (7).

Hart, Kevin. 2013. "Fields of *Dharma*: On T.S. Eliot and Robert Gray." *Literature and Theology* 27 (3): 267–284.

Harvey, William. 1854. "Extract of a Letter from Dr. Harvey, dated Cape Riche, West Australia, March 12, 1854." *Hooker's Journal of Botany and Kew Garden Miscellany* 6: 219.

Haseman, Brad. 2006. "A Manifesto for Performative Research." *Media International Australia* 118: 98–106.

Hasluck, Alexandra. 1955. *Portrait with Background: A Life of Georgiana Molloy*. Melbourne: Oxford University Press.

Hassell, Ethel. 1975. *My Dusky Friends*. Dalkeith, Australia.: C.W. Hassell.

Hayles, N. Katherine. 1999. *How We Became Posthuman: Virtual Bodies in Cybernetics, Literature, and Informatics*. Chicago: The University of Chicago.

Hayles, N. Katherine. 2002. "Flesh and Metal: Reconfiguring the Mindbody in Virtual Environments." *Configurations* 2: 297–320.

Hazell, Dinah. 2006. *The Plants of Middle-earth: Botany and Sub-creation*. Kent, OH: The Kent State University Press.

Hefferon, Kathleen. 2012. *Let Thy Food Be Thy Medicine: Plants and Modern Medicine*. Oxford: Oxford University Press.

Heidegger, Martin. 1982. *On the Way to Language*. Translated by P Hertz. San Francisco, CA: Harper & Row.

Hitchings, Russell. 2003. "People, Plants and Performance: On Actor Network Theory and the Material Pleasures of the Private Garden." *Social & Cultural Geography* 4 (1): 99–113.

Holmes, Oliver Wendell. 1870. *The Poems of Oliver Wendell Holmes*. Boston: Fields, Osgood, & Co.

Homans, Peter. 1989. *The Ability to Mourn: Disillusionment and the Social Origins of Psychoanalysis*. Chicago: University of Chicago Press.

Homestead, William. 2014. "The Language That All Things Speak: Thoreau and the Voice of Nature." In *Voice and Environmental Communication*, edited by Jennifer Peeples and Stephen Depoe, 183–204. Houndsmills: Palgrave Macmillan.

Hope, Cat, and John Ryan. 2014. *Digital Arts: An Introduction to New Media*. London: Bloomsbury.

Hopper, Stephen. 1998. "An Australian Perspective on Plant Conservation Biology Practice." In *Conservation Biology for the Coming Decade*, edited by Peggy Fiedler and Peter Kareiva, 255–278. New York: Chapman Hall.

Hopper, Stephen. 2004. "Southwestern Australia, Cinderella of the World's Temperate Floristic Regions." *Curtis's Botanical Magazine* 21 (2): 132–180.

Hopper, Stephen. 2010. "Nuytsia floribunda." *Curtis's Botanical Magazine* 26 (4): 333–368. doi: 10.1111/j.1467-8748.2009.01671.x.

Hostetler, Mark. 2012. *The Green Leap: A Primer for Conserving Biodiversity in Subdivision Development*. Berkeley, CA: University of California Press.

Hostettman, Kurt, and Andrew Marston. 1986. "Plants Used in African
    Traditional Medicine." In *Folk Medicine: The Art and Science*, edited by
    Richard Steiner, 111–124. Washington: American Chemical Society.

Hutchens, Alma. 1991. *Indian Herbalogy of North America*. Boston: Shambhala.

Jaeger, Peter. 2013. *John Cage and Buddhist Ecopoetics: John Cage and the
    Performance of Nature*. London: Bloomsbury.

James, David. 2009. Interview with David James by John Ryan. Forrestdale,
    Western Australia, September 2009.

Janaway, Christopher. 2003. "Kant's Aesthetics and the 'Empty Cognitive
    Shock'." In *Kant's Critique of the Power of Judgment: Critical Essays*,
    edited by Paul Guyer, 67–86. Lanham, MD: Rowman & Littlefield.

Johnson Gottesfeld, Leslie, and Beverley Anderson. 1988. "Gitksan Traditional
    Medicine: Herbs and Healing." *Journal of Ethnobiology* 8 (1): 13-33.

Johnson, Kent, ed. 1991. *Beneath a Single Moon: Buddhism in Contemporary
    American Poetry*. Boston: Shambhala.

Jones, Rachel Bailey. 2011. *Postcolonial Representations of Women: Critical
    Issues for Education*. Dordrecht: Springer.

Jones, Steven. 2013. *The Emergence of the Digital Humanities*. Hoboken: Taylor
    and Francis.

Kane, Paul. 2010. "East-West Turnings: Australian and American Poetry in Light
    of Asia." In *Reading Across the Pacific: Australia-United States
    Intelelctual Histories*, edited by Robert Dixon and Nicholas Birns, 107–
    118. Sydney, Australia: Sydney University Press.

Kant, Immanuel. 1978. *Anthropology from a Pragmatic Point of View*. Translated
    by Victor Lyle Dowdell. Carbondale, IL: Southern Illinois University
    Press.

Kant, Immanuel. 1997. *The Cambridge Edition of the Works of Immanuel Kant:
    Lectures on Metaphysics*. Edited by Paul Guyer and Allen Wood.
    Cambridge, UK: Cambridge University Press.

Karban, Rick. 2016. "The Language of Plant Communication (and How it
    Compares to Animal Communication)." In *The Language of Plants:
    Science, Philosophy, Literature*, edited by Patricia Vieira, Monica
    Gagliano, and John Ryan. Manunscript in press.

Kew Royal Botanic Gardens. 2014. *Useful Plants and Fungi* Retrieved July 18,
    2014, from, www.kew.org/science-conservation/plants-fungi/useful.

Kiene, Tobias. 2011. *The Legal Protection of Traditional Knowledge in the Pharmaceutical Field.* Münster: Waxmann Verlag.

Kinsella, John. 2011. *Armour.* London: Picador.

Klein, Julie Thompson. 1990. *Interdisciplinarity: History, Theory, and Practice.* Detroit, MI: Wayne State University Press.

Knapp, Sonja. 2010. *Plant Biodiversity in Urbanized Areas.* Wiesbaden, Germany: Springer Fachmedien.

Knickerbocker, Scott. 2012. *Ecopoetics: The Language of Nature, the Nature of Language.* Amherst: University of Massachusetts Press.

Korsmeyer, Carolyn. 1999. *Making Sense of Taste: Food and Philosophy.* Ithaca, NY: Cornell University Press.

Krajewski, Pascal. 2013. *L'art au Risque de la Technologie: Le Glaçage du Sensible.* Vol. 2. Paris: L'Harmattan.

Kreitman, Norman. 2006. The Varieties of Aesthetic Distinterestedness. *Contemporary Aesthetics,* http://www.contempaesthetics.org/newvolume/pages/article.php?articleID=390.

Lambert, Joe. 2013. *Digital Storytelling: Capturing Lives, Creating Community.* Hoboken: Taylor and Francis.

Langford, Martin. 2012. Cumulus: Robert Gray and the Visual Arts of Poetry. *Meanjin,* meanjin.com.au/articles/post/cumulus-robert-gray-and-the-visual-arts-of-poetry.

Lansdown, Andrew. 1979. *Homecoming.* Fremantle: Fremantle Arts Centre Press.

Lasserre, Grégory, and Anaïs Met den Ancxt. 2014. *Akousmaflore: Sensitive and Interactive Musical Plants*Retrieved June 4, 2014, from, www.scenocosme.com/akousmaflore_en.htm.

Latour, Bruno. 1987. *Science in Action: How to Follow Scientists and Engineers Through Society.* Milton Keynes: Open University Press.

Latour, Bruno. 2005. *Reassembling the Social: An Introduction to Actor-Network Theory.* Oxford: Oxford University Press.

Leopold, Aldo. 1993. "The Land Ethic." In *Environmental Philosophy: From Animal Rights to Radical Ecology,* edited by Michael Zimmerman and J. Baird Callicott, 95–109. Englewood Cliffs: Prentice-Hall.

Lessig, Lawrence. 2004. *Free Culture: The Nature and Future of Creativity.* New York: Penguin.

Li, Xiao-Li, Shi Sun, Guang-Jian Du, Lian-Wen Qi, Stanley Williams, Chong-Zhi Wang, and Chun-Su Yuan. 2010. "Effects of *Oplopanax horridus* on Human Colorectal Cancer Cells." *Anticancer Research: International Journal of Cancer Research and Treatment* 30 (2): 295–302.

Lin, Chen-kuo. 1997. "Metaphysics, Suffering and Liberation: The Debate Between Two Buddhisms." In *Pruning the Bodhi Tree: The Storm over Critical Buddhism*, edited by Jaime Hubbard and Paul Swanson, 298–313. Honolulu: University of Hawaii Press.

Lindley, John. 1840. *A Sketch of the Vegetation of the Swan River Colony*. London: James Ridgway.

Lister, Martin, Jon Dovey, Seth Giddings, Iain Grant, and Kieran Kelly. 2003. *New Media: A Critical Introduction*. London: Routledge.

Lixinski, Lucas. 2013. *Intangible Cultural Heritage in International Law*. Oxford: Oxford University Press.

Loh, Jonathan, and David Harmon. 2005. "A Global Index of Biocultural Diversity." *Ecological Indicators* 5: 231–241.

Long, Colin, and Anita Smith. 2010. "Cultural Heritage and the Global Environmental Crisis." In *Heritage and Globalisation*, edited by S Labadi and C Long, 173–191. Milton Park, UK: Routledge.

Lönnrot, Elias. 2007. *Kalevala: The Land of the Heroes, in Two Volumes*. Translated by William Forsell Kirby. New York: Cosimo.

Lovejoy, Margot, Christiane Paul, and Victoria Vesna. 2011. "Introduction." In *Context Providers: Conditions of Meaning in Media Arts*, edited by Margot Lovejoy, Christiane Paul and Victoria Vesna, 1–10. Bristol: Intellect.

Lovelock, James. 1988. *The Ages of Gaia: A Biography of Our Living Earth*. Oxford: Oxford University Press.

Mabey, Richard. 2010. *Weeds: How Vagabond Plants Catecrashed Civilisation and Changed the Way We Think about Nature*. London: Profile Books Ltd. .

MacDonald, Lindsay (ed.). 2006. *Digital Heritage: Applying Digital Imaging to Cultural Heritage*. Amsterdam and London: Elsevier.

Maffi, Luisa. 2001. "Introduction: On the Interdependence of Biological and Cultural Diversity." In *On Biocultural Diversity: Linking Language, Knowledge, and the Environment*, edited by Luisa Maffi, 1–50. Washington and London: Smithsonian Institution Press.

Maffi, Luisa. 2005. "Linguistic, Cultural, and Biological Diversity." *Annual Review of Anthropology* 29: 599–617.

Maffi, Luisa. 2008. "Biocultural Diversity and Sustainability." In *Handbook of Environment and Society*, edited by Jules Pretty, Ted Benton, Julia Guivant, David Lee, David Orr and Max Pfeffer, 267–277. London: SAGE Publications.

Maffi, Luisa, and Ellen Woodley. 2010. *Biocultural Diversity Conservation: A Global Sourcebook*. London: Earthscan.

Main, Barbara York. 1967. *Between Wodjil and Tor*. Brisbane and Perth, Australia: The Jacaranda Press and Landfall Press.

Manes, Christopher. 1996. "Nature and Silence." In *The Ecocriticism Reader: Landmarks in Literary Ecology*, edited by Cheryll Glotfelty and Harold Fromm, 15–29. Athens, GA: The University of Georgia Press.

Maness, L, I Goktepe, H Chen, M Ahmedna, and S Sang. 2014. "Impact of *Phytolacca americana* Extracts on Gene Expression of Colon Cancer Cells." *Phytotherapy Research* 28: 219–223. doi: 10.1002/ptr.4979.

Manoff, Marlene. 2004. "Theories of the Archive from Across the Disciplines." *Portal: Libraries and the Academy* 4 (1): 9–25.

Marder, Michael. 2013a. *Plant-Thinking: A Philosophy of Vegetal Life*. New York: Columbia University Press.

Marder, Michael. 2013b. "What is Plant-Thinking." *Klesis -Revue Philosophique* 25: 124–143.

Marder, Michael. 2016. "To Hear Plants Speak." In *The Language of Plants: Science, Philosophy, Literature*, edited by Patricia Vieira, Monica Gagliano and John Ryan.

Martin, Gary. 2007. *Ethnobotany: A Methods Manual*. London: Earthscan.

Mateer, John. 2010. *The West: Australian Poems 1998–2009*. Fremantle, Australia: Fremantle Press.

Mazzolai, Barbara. 2010. "The Plant as a Biomechatronic System." *Plant Signalling & Behavior* 5 (2): 90–93. doi: 10.4161/psb.5.2.10457.

Mazzolai, Barbara, and Stefano Mancuso. 2013. "Smart Solutions from the Plant Kingdom." *Bioinspiration and Biomimetics* 8 (2): 020301. doi: 10.1088/1748-3182/8/2/020301.

McDowell, Michael. 1996. "The Bakhtinian Road to Ecological Insight." In *The Ecocriticism Reader: Landmarks in Literary Ecology*, edited by Cheryll

Glotfelty and Harold Fromm, 371–391. Athens, GA: The University of Georgia Press.

McKinney, Michael. 2005. "Urbanization as a Major Cause of Biotic Homogenization." *Biological Conservation* 127 (3): 247–260.

McNiece, Ray, and Larry Smith, eds. 2004. *America Zen: A Gathering of Poets*. Huron, OH: Bottom Dog Press.

Millenium Ecosystem Assessment. 2005. *Ecosystems and Human Well-Being: Scenarios*. Washington, DC: Island Press.

Miller, Joseph. 1722. *Botanicum Officinale; Or a Compendious Herbal*. London: E. Bell, J. Senex, W. Taylor and J. Osborn.

Millett, Janet. 1980. *An Australian Parsonage or, the Settler and the Savage in Western Australia*. Nedlands: University of Western Australia Press.

Moerman, Daniel. 1998. *Native American Ethnobotany*. Portland: Timber Press.

Mollier, Christine. 2008. *Buddhism and Taoism Face to Face: Scripture, Ritual, and Iconographic Exchange in Medieval China*. Honolulu: University of Hawaii Press.

Montopoli, Monica, Riccardo Bertin, Zheng Chen, Jenny Bolcato, Laura Caparrotta, and Guglielmina Froidi. 2012. "*Croton lechleri* Sap and Isolated Alkaloid Taspine Exhibit Inhibition Against Human Melanoma SK23 and Colon Cancer HT29 Cell Lines." *Journal of Ethnopharmacology* 144 (3): 747–753.

Moore, George Fletcher. 1846. *Descriptive Vocabulary of the Language in Common Use Amongst the Aborigines of Western Australia*. Nedlands, Australia: University of Western Australia Press. Reprint, 1978.

Moore, George Fletcher. 1978. *Diary of Ten Years Eventful Life of an Early Settler in Western Australia and also A Descriptive Vocabulary of the Language of the Aborigines*. Nedlands, Australia: University of Western Australia Press.

Moran, Joe. 2010. *Interdisciplinarity*. 2 ed. Abington, UK: Routledge.

Morris, Daniel. 2006. *The Poetry of Louise Glück: A Thematic Introduction*. Columbia: University of Missouri Press.

Morton, Tim. 2010. "The Dark Ecology of Elegy." In *The Oxford Handbook of the Elegy*, edited by Karen Weisman, 251–271. Oxford: Oxford University Press.

Mules, Warwick. 2006. "Contact Aesthetics and Digital Arts: At the Threshold of the Earth." *Fibreculture* 9: unpaginated.

Mules, Warwick. 2014. *With Nature: Nature Philosophy as Poetics through Schelling, Heidegger, Benjamin and Nancy*. Bristol and Chicago: Intellect Press.

Murray, John. 1995. "Preface." In *Nature Writing Handbook: A Creative Guide*. San Francisco, CA: Sierra Club Books.

Murray, Les. 2007. *Selected Poems*. Melbourne: Black Inc.

Murray, Michael. 2013. "Botanical Medicine: A Modern Perspective." In *Textbook of Natural Medicine*, edited by Joseph Pizzorno and Michael Murray, 255–260. St Louis: Churchill Livingstone.

Myers, Natasha. 2013. Plant Vocalities: Articulating Botanical Sensoria in the Experimental Arts and Sciences. Unpublished Presentation, Sydney, Australia.

Nannup, Noel. 2010. Unpublished Interview with Noel Nannup by John Ryan. Mount Lawley, Western Australia, July 2010.

National Library of Australia. 2014. *Trove* Retrieved April 26, 2014, from, http://trove.nla.gov.au/.

Nemitz, Barbara. 2000. *Trans Plant: Living Vegetation in Contemporary Art*. Ostfildern-Ruit, Germany: Hatje Cantz Publishers.

Nicolescu, Basarab. 2002. *Manifesto of Transdisciplinarity*. Translated by Karen-Claire Voss. Albany, NY: State University of New York Press.

Niemeier, Susanne. 2011. "Culture-Specific Concepts of Emotionality and Rationality." In *Bi-Directionality in the Cognitive Sciences: Avenues, Challenges, and Limitations*, edited by Marcus Callies, Wolfram Keller and Astrid Lohöfer, 43–56. Amsterdam: John Benjamins Publishing Company.

Noble, Robert. 1990. "The Discovery of the Vinca Alkaloids: Chemotherapeutic Agents Against Cancer." *Biochemistry and Cell Biology* 68 (12): 1344–1351.

North, Marianne. 2011. *Recollections of a Happy Life, Volume 2*. Edited by Jane Symonds. Cambridge: Cambridge University Press.

O'Rourke, Karen. 2013. *Walking and Mapping: Artists as Cartographers*. Cambridge, MA: MIT Press.

Olwig, Kenneth. 2006. "Introduction." In *The Nature of Cultural Heritage, and the Culture of Natural Heritage--Northern Perspectives on a Contested Patrimony*, edited by Kenneth Olwig and David Lowenthal, 1–5. Abington: Routledge.

Oosterveer, Peter, and David Sonnenfeld. 2012. *Food, Globalization and Sustainability*. Milton Park: Routledge.

Oskin, Becky. 2013. *Sound Garden: Can Plants Actually Talk and Hear?* Retrieved March 11, 2015, from, http://www.livescience.com/27802-plants-trees-talk-with-sound.html.

Ovid. 2010. *Metamorphoses*. Translated by Stanley Lombardo. Indianapolis: Hackett Publishing Company.

Paczkowska, Grazyna. 2013. *Boronia megastigma Bartl.: Scented Boronia* Retrieved April 24, 2013, from, http://florabase.dec.wa.gov.au/browse/profile/4428.

Papayannis, Thymio, and Peter Howard. 2013. "Editorial: Nature as Heritage." In *Natural Heritage: At the Interface of Nature and Culture*, edited by Peter Howard and Thymio Papayannis, ix–xviii. Hoboken: Taylor and Francis.

Parker, Gilbert. 1892. *Round the Compass in Australia*. London: Hutchinson and Co.

Parker, Pat. 1962. "Boronia Farm." *The Australian Women's Weekly* 24, October 4.

Parry, Ross. 2010. "The Practice of Digital Heritage and the Heritage of Digital Practice." In *Museums in a Digital Age*, edited by Ross Parry, 1–7. Milton Park, UK: Routledge.

Patel, Narayan. 1986. "Ayurveda: The Traditional Medicine of India." In *Folk Medicine: The Art and Science*, edited by Richard Steiner, 41–66. Washington: American Chemical Society.

Paul, Christiane. 2003. *Digital Art*. London: Thames & Hudson.

Pausanias. 2012. *Description of Greece, Vol 1*. Translated by James George Frazer. Edited by James George Frazer. Cambridge, UK: Cambridge University Press.

Peeples, Jennifer, and Stephen Depoe. 2014. "Introduction: Voice and the Environment - Critical Perspectives." In *Voice and Environmental Communication*, edited by Jennifer Peeples and Stephen Depoe, 1–17. Palgrave Macmillan.

Pelloe, Emily. 1926. "The Christmas Tree: A Plea for Preservation." *The West Australian* December 27: 5.

Perth Biodiversity Project. n.d. *What is the Perth Biodiversity Project?* West Perth, Australia: Perth Biodiversity Project.

Phillips, Dana. 2006. Thoreau's Aesthetics and 'The Domain of the Superlative'. *Environmental Values* 15 (3): 293–305, http://www.environmentandsociety.org/node/5967.

Plattner, Hasso. 2012. "Preface." In *Design Thinking Research: Measuring Performance in Context*, edited by Hasso Plattner, Christoph Meinel and Larry Leifer, v–vi. Berlin: Springer-Verlag.

Plec, Emily. 2013. "Perspectives on Human-Animal Communication." In *Perspectives on Human-animal Communication: Internatural communication*, edited by Emily Plec, 1–13. London: Routledge.

Pollan, Michael. 2013. "The Intelligent Plant: Scientists Debate a New Way of Understanding Flora." *The New Yorker* December 23 & 30: 92–105.

Pollard, James. 1947. "Boronia." *The West Australian*Saturday 19 July:4.

Porteous, Douglas. 1989. *Planned to Death: The Annihilation of a Place Called Howdendyke*. Manchester: Manchester University Press.

Posey, Darrell. 2002. "Commodification of the Sacred through Intellectual Property Rights." *Journal of Ethnopharmacology* 83 (1–2): 3–12.

Powers, John. 2007. *Introduction to Tibetan Buddhism*. Ithaca, NY: Snow Lion Publications.

Prelinger, Rick. 2009. "Points of Origin: Discovering Ourselves Through Access." *The Moving Image* 9 (2): 164–175.

Puntarigvivat, Tavivat. 2003. "Buddhadāsa Bhikkhu and Dhammic Socialism." *The Chulalongkorn Journal of Buddhist Studies* 2 (3): 189–207.

Rackham, Oliver. 1996. *Trees and Woodland in the British Landscape: The Complete History of Britain's Trees, Woods, and Hedgerows*. London: Phoenix Press.

Raguso, Robert, and André Kessler. 2016. "The Linguistics of the Plant Headspace." In *The Language of Plants: Science, Philosophy, Literature*, edited by Patricia Vieira, Monica Gagliano and John Ryan.

Ramsay, Stephen. 2011a. *On Building* Retrieved April 9, 2014, from, www.stephenramsay.us/text/2011/01/11/on-building.

Ramsay, Stephen. 2011b. *Who's In and Who's Out* Retrieved April 9, 2014, from, www.stephenramsay.us/text/2011/01/08/whos-in-and-whos-out/.

Reed, Sampson. 1859. *Observations on the Growth of the Mind*. Boston: Crosby, Nichols, and Company, and George Phinney.

Refsland, Scott, Marc Tuters, and Jim Cooley. 2007. "Geo-Storytelling: A Living Archive of Spatial Culture." In *Theorizing Digital Cultural Heritage: A*

Critical Discourse, edited by Fiona Cameron and Sarah Kenderdine, 409–416. Cambridge: MIT Press.

Regan, Tom. 1983. *The Case for Animal Rights*. Berkeley: University of California Press.

Regan, Tom. 1993. "Animal Rights, Human Wrongs." In *Environmental Philosophy: From Animal Rights to Radical Ecology*, edited by Michael Zimmerman and J. Baird Callicott, 33–48. Englewood Cliffs: Prentice-Hall.

Repko, Allen. 2008. *Interdisciplinary Research: Process and Theory*. Los Angeles: Sage.

Rigby, Catherine. 2004. "Earth, World, Text: On the (Im)possibility of Ecopoiesis." *New Literary History* 35 (3): 427–442.

Rivers of Emotion. 2012. *Rivers of Emotion* Retrieved April 27, 2014, from, http://www.riversofemotion.org.au/.

Robinson, Daniel. 2010. *Confronting Biopiracy: Challenges, Cases and International Debates*. London: Earthscan.

Roe, John Septimus. 2005. "Journal of an Expedition from Swan River Overland to King George's Sound." In *Western Australian Exploration, 1826-1835*, edited by Joanne Shoobert, 456–497. Carlisle, WA: Hesperian Press.

Rogers, Richard. 1998. "Overcoming the Objectification of Nature in Constitutive Theories: Toward a Transhuman, Materialist Theory of Communication." *Western Journal of Communication* 62 (3): 244–272.

Rose, Deborah Bird. 1996. *Nourishing Terrains: Australian Aboriginal Views of Landscape and Wilderness*. Canberra: Australian Heritage Commission.

Rose, Deborah Bird. 2008. "Judas Work: Four Modes of Sorrow." *Environmental Philosophy* 5 (2): 51–66.

Rose, Deborah Bird. 2011a. "Flying Foxes: Kin, Keystone, Kontaminant." *Australian Humanities Review* (50).

Rose, Deborah Bird. 2011b. *Wild Dog Dreaming: Love and Extinction*. Charlottesville: University of Virginia Press.

Rose, Deborah Bird, and Libby Robin. 2004. The Ecological Humanities in Action: An Invitation. *Australian Humanities Review* (31–32), http://www.australianhumanitiesreview.org/archive/Issue-April-2004/rose.html.

Rowe, Noel, and Vivian Smith, eds. 2006. *Windchimes: Asia in Australian Poetry*. Canberra: Pandanus Books.

Ruskin, John. 1998. "Of the Pathetic Fallacy." In *The Genius of John Ruskin: Selections from His Writings*, edited by John Rosenberg, 61–72. Charlottesville, VA: University of Virginia Press.

Ryan, John. 2012a. *Green Sense: The Aesthetics of Plants, Place and Language*. Oxford: TrueHeart Press.

Ryan, John. 2012b. *Two With Nature: Botanical Illustrations by Ellen Hickman and Botanical Poetry by John Ryan*. Fremantle: Fremantle Press.

Ryan, John. 2013. *Unbraided Lines: Essays in Environmental Thinking and Writing*. Champaign, IL: Common Ground Publishing.

Ryan, John. 2014. *Being With: Essays in Poetics, Ecology and the Senses*. Champaign, IL: Common Ground Publishing.

Saito, Yuriko. 1998. "Appreciating Nature on Its Own Terms." *Environmental Ethics* 20: 135–149.

Sanderson, Eric. 2009. *Mannahatta: A Natural History of New York City*. New York: Harry N. Abrams, Inc.

Sasaki, T., K. Yamazaki, T. Yamori, and T. Endo. 2002. "Inhibition of Proliferation and Induction of Differentiation of Glioma Cells with Datura Stramonium Agglutinin." *British Journal of Cancer* 87: 918–923.

Saunders, Gill. 1995. *Picturing Plants: An Analytical History of Botanical Illustration*. Berkeley, CA: University of California Press.

Schachtel, Ernest. 1984. *Metamorphosis: On the Development of Affect, Perception, Attention, and Memory*. New York: Da Capo Press.

Schelling, Andrew. 2005. "Preface." In *The Wisdom Anthology of North American Buddhist Poetry*, edited by Andrew Schelling, xiii–xviii. Somerville, MA: Wisdom Publications.

Schmelzer, Gaby. 2008. *Plant Resources of Tropical Africa: Medicinal Plants*. Edited by Gaby Schmelzer. Wageningen: PROTO Foundation and Backhuys Publishers.

Schneider, Stephen Henry. 2004. *Scientists Debate Gaia: The Next Century*. Edited by Stephen Henry Schneider. Cambridge: MIT Press.

Scholte, Jan Aart. 2005. *Globalization: A Critical Introduction*. 2nd ed. Basingstoke: Palgrave MacMillan.

Schultes, Richard Evan, Albert Hoffman, and Christian Rätsch. 2001. *Plants of the Gods: Their Sacred, Healing, and Hallucinogenic Powers*. Rochester: Healing Arts Press.

Searles, Nalda. 2014. Unpublished Interview with Nalda Searles by John Ryan. Midvale, Western Australia, April 2014.

Séquin, Margareta. 2012. *The Chemistry of Plants: Perfumes, Pigments, and Poisons*. Cambridge, UK: The Royal Society of Chemistry.

Serres, Michel. 1995. *The Natural Contract*. Translated by Elizabeth MacArthur and William Paulson. Ann Arbor: The University of Michigan Press.

Serres, Michel. 2008. *The Five Senses: A Philosophy of Mingled Bodies*. London: Continuum.

Serventy, Vincent. 1970. *Dryandra: The Story of an Australian Forest*. Sydney: A.H. and A.W. Reed.

Shearer, B.L, C.E. Crane, S. Barrett, and A. Cochraine. 2007. "*Phytophthora cinnamomi* Invasion, a Major Threatening Process to Conservation of Flora Diversity in the South-west Botanical Province of Western Australia." *Australian Journal of Botany* 55 (3): 225–238.

Shiva, Vandana. 1997. *Biopiracy: The Plunder of Nature and Knowledge*. New York: South End Press.

Skinner, Jonathan. 2015. Ecopoetics. *Jacket 2*, jacket2.org/commentary/jonathan-skinner.

Sliva, Daniel. 2003. "*Ganoderma lucidum* (Reishi) in Cancer Treatment." *Integrative Cancer Therapies* 2 (4): 358–364.

Smil, Vaclav. 2008. *Energy in Nature and Society: General Energetics of Complex Systems*. Cambridge, MA: Massachusetts Institute of Technology.

Smith, David Nowell. 2012. "Scaping the Land: The New British Pastoral." *Chicago Review* 57: 182–193.

Smith, James Edward. 1793. *A Specimen of the Botany of New Holland*. London: J. Sowerby.

Smith, Laurajane. 2006. *Uses of Heritage*. New York: Routledge.

Smith, Mick. 2013. Ecological Community, the Sense of the World, and Senseless Extinction. *Environmental Humanities* 2: 21–41, http://environmentalhumanities.org/arch/vol2/2.2.pdf.

Smith, Thomas Michael, and Robert Leo Smith. 2012. *Elements of Ecology*. 8 ed. San Francisco: Pearson Benjamin Cummings.

Snodgrass, Mary Ellen. 2010. *Encyclopedia of the Literature of Empire*. New York: Infobase Publishing.

Snyder, Gary. 1969. *Earth House Hold: Technical Notes & Queries to Fellow Dharma Revolutionaries*. New York: New Directions Books.

Sommerer, Christa, and Laurent Mignonneau. 2011. "Cultural Interfaces: Interaction Revisited." In *Imagery in the 21st Century*, edited by Oliver Grau and Thomas Veigl, 201–218. Cambridge, MA: MIT Press.

Soulé, Michael , and Daniel Press. 1998. "What is Environmental Studies?" *BioScience* 48 (5): 397–405.

Sowards, Adam. 2007. *United States West Coast: An Environmental History*. Santa Barbara: ABC-CLIO.

Spooner, Rosie. 2010. *Kunstkammer/Wunderkammer: 10th Annual InterAccess Emerging Artists Exhibition*Retrieved June 5, 2014, from,www.singingplants.org/uploads/4/3/9/8/4398299/w_essay.pdf.

Steffan, Will, Andrew Burbidge, Lesley Hughes, Roger Kitching, David Lindenmayer, Warren Musgrave, Mark Stafford Smith, and Patricia Werner. 2009. *Australia's Biodiversity and Climate Change*. Collingwood: CSIRO Publishing.

Stow, Randolph. 1969. *A Counterfeit Silence: Selected Poems*. Sydney: Angus and Robertson.

Stow, Randolph. 2009. "The Land's Meaning." In *The Penguin Anthology of Australian Poetry*, edited by John Kinsella, 248. London: Penguin.

Sunday Times. 1940. "Christmas Trees Cut Down to Make Dart Boards." *Sunday Times* December 1:7.

Svoboda, Robert. 1992. *Ayurveda: Life, Health, and Harmony*. London and New York: Arkana.

Tamokou, Jean de Dieu, Jean Rodolphe Chouna, Eva Fischer-Fodor, Gabriela Chereches, Otilia Barbos, Grigore Damian, Daniela Benedec, Mihaela Duma, Alango Pépin Nkeng Efouet, and Hippolyte Kamdem Wabo. 2013. "Anticancer and Antimicrobial Activities of Some Antioxidant-Rich Cameroonian Medicinal Plants." *PLOS One*8 (2): 1–14. doi: 10.1371/journal.pone.0055880.

Tan, Joan Qionglin. 2009. *Han Shan, Chan Buddhism and Gary Snyder's Ecopoetic Way*. Eastbourne, UK: Sussex Academic Press.

Tantaquidgeon, Gladys. 1972. *Folk Medicine of the Delaware and Related Algonkian Indians*. Harrisburg: Pennsylvania Historical and Museum Commission.

Tarlo, Harriet. 2011. "Introduction." In *The Ground Aslant: An Anthology of Radical Landscape*, edited by Harriet Tarlo, 7–18. Exeter: Shearsman Books.

The Free Dictionary. 2014. *Archive* Retrieved April 25, 2014, from, http://www.thefreedictionary.com/archive.

The Huffington Post UK. 2013. Skin Cancer Treatment: Liquorice Root May Fight Malignant Melanoma. *The Huffington Post*, http://www.huffingtonpost.co.uk/2013/10/10/liquorice-root-may-fight-skin-cancer_n_4075898.html.

The West Australian. 1919. "The Christmas Tree: Its Parasitic Nature." *The West Australian*April 16:6.

The West Australian. 1933. "A Presidential Chair: Gift to British Pharmacists." *The West Australian*.

The Western Australian Herbarium. 2014. *FloraBase: The Western Australian Flora* Retrieved April 7, 2014, from, https://florabase.dpaw.wa.gov.au.

Thoreau, Henry David. 1993. *Faith in a Seed: The Dispersion of Seeds and Other Late Natural History Writings*. Washington, DC: Island Press.

Thoreau, Henry David. 2000. *Wild Fruits: Thoreau's Rediscovered Last Manuscript*. Edited by Bradley P. Dean. New York: W. W. Norton.

Thoreau, Henry David. 2004. *Walden*. London: CRW Publishing Limited.

Thoreau, Henry David. 2005. *Civil Disobedience and Other Essays*. Stilwell, KS: Digireads.com Publishing.

Thoreau, Henry David. 2007. *Walking*. LaVergne, TN: Filiquarian Publishing.

Thoreau, Henry David. 2009. *The Journal of Henry David Thoreau, 1837–1861*. New York: The New York Review of Books.

Tindle, Hilary, Roger Davis, Russell Phillips, and David Eisenberg. 2005. "Trends in the Use of Complementary and Alternative Medicine by US Adults: 1997–2002." *Alternative Therapies in Health and Medicine* 11: 42–49.

Todd, Cain. 2010. *The Philosophy of Wine: A Case of Truth, Beauty and Intoxication*. New York: Routledge.

Tolkien, J.R.R. 1964. *Tree and Leaf*. London: Unwin Books.

Tolkien, J.R.R. 1977. *The Silmarillion*. Edited by Christopher Tolkien. London: George Allen & Unwin.

Tolkien, J.R.R. 1980. *Unfinished Tales of Númenor and Middle-earth*. London: George Allen & Unwin.

Tolkien, J.R.R. 1981. *Letters of J. R. R. Tolkien*. Edited by Humphrey Carpenter and Christopher Tolkien. London: George Allen & Unwin Ltd.

Tolkien, J. R.R. 1985. *The History of Middle-earth: The Lays of Beleriand*. Edited by Christopher Tolkien. Boston: Houghton Mifflin.

Tolkien, J.R.R. 1990. *The History of Middle-earth: The War of the Ring*. Boston: Houghton Mifflin.

Tolkien, J.R.R. 1994a. *The Return of the King: Being the Third Part of The Lord of the Rings*. New York: Houghton Mifflin Company.

Tolkien, J.R.R. 1994b. *The Fellowship of the Ring: Being the First Part of The Lord of the Rings*. New York: Houghton Mifflin Harcourt Publishing.

Tolkien, J.R.R. 1994c. *The Two Towers: Being the Second Part of The Lord of the Rings*. New York: Houghton Mifflin Harcourt Publishing.

Tolkien, J. R. R. 1995. *The Hobbit*. Boston: Houghton Mifflin Harcourt.

Trewavas, Anthony. 2002. "Mindless Mastery." *Nature* 415: 841.

Trewavas, Anthony. 2003. "Aspects of Plant Intelligence." *Annals of Botany* 92 (1):1–20. doi: 10.1093/aob/mcg101.

Trewavas, Anthony. 2006. "The Green Plant as an Intelligent Organism." In *Communication in Plants: Neuronal Aspects of Plant Life*, edited by František Baluška, Stefano Mancuso and Dieter Volkmann, 1–18. Berlin: Springer-Verlag.

Trewavas, Anthony. 2014. *Plant Behaviour and Intelligence*. Oxford: Oxford University Press.

Trigger, David, and Jane Mulcock. 2005. "Forests as Spiritually Significant Places: Nature, Culture and "Belonging" in Australia." *The Australian Journal of Anthropology* 16 (3): 306–320.

Trigilio, Tony. 2007. *Allen Ginsberg's Buddhist Poetics*. Carbondale, IL: Southern Illinois University Press.

Troy, Timothy. 1990. "*Ktaadn*: Thoreau the Anthropologist." *Dialectical Anthropology* 15 (1): 74–81.

Tschida, David. 2014. "The Ethics of Listening in the Wilderness Writings of Sigurd F. Olsen." In *Voice and Environmental Communication*, edited by Jennifer Peeples and Stephen Depoe, 205–227. Palgrave Macmillan.

Tsing, Anna. 2010. "Worlding the Matsutake Diaspora or, Can Actor-Network
        Theory Experiment with Holism?" In *Experiments in Holism*, edited by
        Nils Bubandt and Ton Otto, 47–66. London, England: Wiley-Blackwell.

Tuan, Yi-fu. 1974. *Topophilia: A Study of Environmental Perception, Attitudes,
        and Values*. Englewood Cliffs: Prentice-Hall.

Tyler, J.E.A. 1979. *The New Tolkien Companion*. New York: St. Martin's Press.

UbuWeb. 2014. *UbuWeb* Retrieved April 26, 2014, from,
        http://www.ubuweb.com/.

University of Michigan. 2014. *Native American Ethnobotany* Retrieved July 18,
        2014, from, herb.umd.umich.edu.

Van Dooren, Thom. 2010. "Pain of Extinction: The Death of a Vulture." *Cultural
        Studies Review* 16 (2): 271–89.

Van Dooren, Thom. 2011. Vultures and their People in India: Equity and
        Entanglement in a Time of Extinctions. *Australian Humanities Review*
        50: 45–61, http://www.australianhumanitiesreview.org/archive/Issue-
        May-2011/vandooren.html.

Van Vuuren, Detlef, Osvaldo Sala, and Henrique Pereira. 2006. The Future of
        Vascular Plant Diversity under Four Global Scenarios. *Ecology and
        Society* 11 (2), http://www.ecologyandsociety.org/vol11/iss2/art25.

Varner, Gary. 2006. *The Mythic Forest, the Green Man and the Spirit of Nature: .*
        New York: Algora Publishing.

Vidal, Fernando. 2011. *Endangerment and Its Consequences: Documenting and
        Preserving Nature and Culture* Retrieved September 28, 2014, from,
        http://www.mpiwg-berlin.mpg.de/en/news/features/feature24

Vieira, Patricia, Monica Gagliano, and John Ryan, eds. 2016. *The Language of
        Plants: Science, Philosophy, Literature*. New York: Columbia
        University Press.

Vogel, Virgil. 1970. *American Indian Medicine*. Norman: University of
        Oklahoma Press.

Waldstein, Anna. 2014. "How Can Ethnobotany Contribute to the History of
        Western Medicine? A Mesoamerican Answer." In *Critical Approaches
        to the History of Western Herbal Medicine: From Classical Antiquity to
        the Early Modern Period*, edited by Susan Francia and Anne Stobart,
        271–288. London: Bloomsbury.

Walker, Steve. 2010. *The Power of Tolkien's Prose: Middle-Earth's Magical
        Style*. Houndsmill: Palgrave Macmillan.

Walls, Laura Dassow. 1995. *Seeing New Worlds: Henry David Thoreau and Nineteenth-Century Natural Science*. Madison, WI: University of Wisconsin Press.

Wardell, Lois, and Charlotte Rowe. 2010. *Cactus Acoustics* Retrieved April 11, 2015, from, http://arapahost.com/selected_science_projects/phyto-acoustics.

Watts, D.C. 2007. *Elsevier's Dictionary of Plant Lore*. Burlington, MA: Academic Press.

Watts, E.K. 2001. "'Voice' and 'Voicelessness' in Rhetorical Studies." *Quarterly Journal of Speech* 87: 179–196.

Watts, Eric King. 2013. "Coda: Food, Future, Zombies." In *Voice and Environmental Communication*, edited by Jennifer Peeples and Stephen Depoe, 257–263. Palgrave Macmillan.

Weldon, Annamaria. 2014. *The Lake's Apprentice*. Crawley, WA: University of Western Australia Publishing.

Western Mail. 1936. "Boronia: Story of Local Perfumery Industry." *Western Mail*Thursday, October 1:22.

Whalen-Bridge, John, and Gary Storhoff, eds. 2009. *The Emergence of Buddhist American Literature*. Albany: SUNY Press.

Willander, Johan, and Maria Larsson. 2006. "Smell Your Way Back to Childhood: Autobiographical Odor Memory." *Psychonomic Bulletin & Review*13 (2):240–244.

Williams, Nerys. 2011. *Contemporary Poetry*. Edinburgh: Edinburgh University Press.

Williams, Raymond. 1982. *Socialism and Ecology*. London: Socialist Environment and Resources Association.

Willson, Lawrence. 1959. "Thoreau: Student of Anthropology." *American Anthropologist* 61 (2): 279–289.

Wilson, Edward O. 1986. *Biophilia*. Cambridge: Harvard University Press.

Wilson, Edward O. 1998. *Consilience: The Unity of Knowledge*. New York: Alfred A. Knopf.

Wilson, Stephen. 2002. *Information Arts: Intersections of Art, Science, and Technology*. Cambridge, MA: The MIT Press.

Wolfe, Cary. 2010. *What Is Posthumanism?* Minneapolis: University of Minnesota Press.

Wood, Thomas. 1943. *Cobbers: A Personal Record of a Journey from Essex, in England, to Australia, Tasmania and Some of the Reefs and Islands in the Coral Sea, Made in the Years 1930, 1931 and 1932*. Melbourne: Oxford University Press.

Worster, Donald. 1994. *Nature's Economy: A History of Ecological Ideas*. Cambridge, UK: Cambridge University Press.

Wright, Judith. 1994. *Judith Wright: Collected Poems 1942–1985*. Sydney: Angus & Robertson.

Yance, Donald. 2013. *Adaptogens in Medical Herbalism: Elite Herbs and Natural Compounds for Mastering Stress, Aging, and Chronic Disease*. Rochester: Healing Arts Press.

Young, Peter. 2013. *Oak*. London: Reaktion Books.

Yusoff, Kathryn. 2011. "Aesthetics of Loss: Biodiversity, Banal Violence and Biotic Subjects." *Transactions of the Institute of British Geographers* 37: 578–592.

Zolberg, Vera. 2003. "'An Elite Experience for Everyone': Art Museums, the Public, and Cultural Literacy." In *Museum Culture: Histories, Discourses, Spectacles*, edited by Daniel Sherman and Irit Rogoff, 49–65. London: Taylor & Francis.

ABOUT THE AUTHOR

John Charles Ryan is a postdoctoral research fellow in Communications and Arts at Edith Cowan University in Australia. He is the author of *Green Sense* (2012), *Two with Nature* (2012, with Ellen Hickman), *Unbraided Lines* (2013), *Digital Arts* (2014, with Cat Hope), and *Being With* (2014). He is the co-editor of two forthcoming collections in the field of critical plant studies: *Green Thread* (2015) and *The Language of Plants* (2016) with Patrícia Vieira and Monica Gagliano. His interests include the environmental humanities, ecocriticism, ecocultural studies, ecopoetics, and practice-led research. His project FloraCultures is a digital archive of plant-based cultural heritage (www.floracultures.org.au).

www.ingramcontent.com/pod-product-compliance
Lightning Source LLC
Chambersburg PA
CBHW032131020426
42334CB00016B/1122